SpringerBriefs in Electrical and Computer Engineering

Cooperating Objects

Series Editor

Pedro José Marron, Duisburg, Germany

For further volumes:
http://www.springer.com/series/10208

Jose Ramiro Martinez-de Dios
Adrian Jimenez-Gonzalez
Alberto de San Bernabe
Anibal Ollero

A Remote Integrated Testbed for Cooperating Objects

 Springer

Jose Ramiro Martinez-de Dios
Adrian Jimenez-Gonzalez
Alberto de San Bernabe
Anibal Ollero
Universidad de Sevilla
Seville
Spain

ISSN 2191-8112   ISSN 2191-8120   (electronic)
ISBN 978-3-319-01371-8   ISBN 978-3-319-01372-5   (eBook)
DOI 10.1007/978-3-319-01372-5
Springer Cham Heidelberg New York Dordrecht London

Library of Congress Control Number: 2013944205

Printed on acid-free paper

Springer is part of Springer Science+Business Media (www.springer.com)

To our families,
To our friends

Preface

This book describes the development and validation of the CONET Cooperating Objects Integrated Testbed. The domain of Cooperating Objects is a cross-section between networked robots, ubiquitous computing, and (wireless) sensor networks. Existing tools that have been developed for these individual fields cannot be used to research in a domain that largely surpasses beyond these fields frontiers. New testbeds are required to support the experimental assessment and evaluation of Cooperating Objects techniques and algorithms.

Interoperability between elements from different technological fields and (broadly understood) heterogeneity are central topics in Cooperating Objects. Both have been the main concepts in the development of the testbed described in this book. The CONET Cooperating Objects Integrated Testbed has been designed to allow full equanimity and interoperability between heterogeneous elements from different technological fields giving the possibility to cover an unprecedented range of experiments involving cooperation among mobile robots and sensor networks.

The presented testbed has been developed within the Cooperating Objects Network of Excellence (CONET) (http://www.cooperating-objects.eu) co-funded by the European Commission with the aim to identify and promote work on the main research topics in Cooperating Objects in the short, medium and long terms. The main objective of the CONET Integrated Testbed is to become a benchmark to facilitate comparison and assessment of Cooperating Objects techniques and algorithms from academic and industrial communities.

We hope that this book may spark new improvements and innovative ideas among the growing community of testbed users and designers.

Seville, June 2013

Jose Ramiro Martinez-de Dios
Adrian Jimenez-Gonzalez
Alberto de San Bernabe
Anibal Ollero

Acknowledgments

The authors thank all members of the Robotics, Vision, and Control Group at the University of Seville for their contribution and support in the development of the testbed described in this book. The authors are grateful to Mr. Gabriel Nunez, Mr. Jose Manuel Sanchez-Matamoros, Mr. Karim Lferd, Mr. Arturo Torres, Prof. Fernando Caballero, and Mr. Victor Vega, for their contribution and support in the implementation of the testbed.

The authors would like to thank the School of Engineering of the University of Seville (http://www.esi.us.es) and particularly to Mr. Pedro Arco for supporting the testbed and providing access and permissions for the use of the testbed room.

We would also like to express our gratitude to all of the members of the CONET consortium. Their insight comments, suggestions, support, and feedback have been crucial to improve the testbed.

Contents

Acronyms

API	Application Programming Interface
BAN	Body Area Network
CH	Cluster Head
CO	Cooperating Object
CONET	Cooperating Objects Network of Excellence
COTS	Commercial Off-the-Shelf
EKF	Extended Kalman Filter
GPS	Global Positioning System
GUI	Graphical User Interface
IMU	Inertial Measurement Unit
LAN	Local Area Network
LQI	Link Quality Indicator
OS	Operating System
PDA	Personal Digital Assistant
PF	Particle Filter
R/C	Radio Controlled
RGB	Red, Green, Blue
RGB-D	Red, Green, Blue, Distance
RSSI	Received Signal Strength Indicator
SLAM	Simultaneous Localization and Mapping
UAV	Unmanned Aerial Vehicle
USAR	Urban Search and Rescue
VPN	Virtual Private Network
WSN	Wireless Sensor Networks

Chapter 1
Introduction

1.1 Cooperating Objects

In the last years different technological fields have emerged in the broad context of embedded systems. Disciplines such as pervasive and ubiquitous computing, where objects of everyday use are endowed with sensing, computational and communication capabilities, have appeared. Also, nodes in Wireless Sensor Networks (WSN) can collaborate together using ad-hoc network technologies to achieve a common mission of supervision of some area or some particular process using a set of low-cost sensors. The synergies among these technologies, and others such as robotics, have facilitated their convergence in what has been called Cooperating Objects.

As defined in [1]:

> "Cooperating Objects are modular systems of autonomous, heterogeneous devices pursuing a common goal by cooperation in computations and in sensing and/or actuating with the environment."

The domain of Cooperating Objects is a cross-section between networked embedded systems, ubiquitous computing and (wireless) sensor networks. Cooperation and interoperability between elements from different technological fields play a central role. Cooperation among heterogeneous systems is not easy. Devices such as mobile robots and WSN nodes present high levels of diversity in its sensing, computational and communication capabilities. Sensors used in each of these platforms also have high degree of heterogeneity in technology, sensing features, output bandwidth, interfaces and power consumption, among others. However, it is this heterogeneity and the potential synergies it can originate what makes cooperation-based approaches useful for a wide range of problems.

1.2 Role of Testbeds in Cooperating Objects Research

A new set of tools are required to support research in this emerging Cooperating Objects domain. Architectures, middleware, testbeds, simulators that have been developed for individual fields cannot be used to research in approaches or tech-

J. R. Martinez-de Dios et al., *A Remote Integrated Testbed for Cooperating Objects*, SpringerBriefs in Cooperating Objects, DOI: 10.1007/978-3-319-01372-5_1, © The Author(s) 2014

niques that surpass beyond its field constraints. This book deals with testbeds for Cooperating Objects research.

In the last years experimentation in fields such as multi-robot systems and sensor networks has originated a growing demand of testbeds. In some fields validation with data sets or simulations is widely assumed. However, the complexity of multi-object cooperation requires experimentation in testbeds since they provide a degree of realism that can be hardly obtained with simulations.

Despite the high number and variety of testbeds for individual fields such as robotics and WSN, those that provide high cross-interoperability between different Cooperating Objects fields are still very scarce. In fact, the lack of suitable tools for testing and validating algorithms involving heterogenous devices, technologies and techniques from different technological fields has been identified as one of the main difficulties in Cooperating Objects research [2, 3].

Testbeds have two main objectives. On one hand, they should be developed to provide a controlled environment to allow algorithm testing and debugging in real hardware with simulation-like conditions. On the other hand, they should also be suitable to fill the gap between research and market, allowing to test methods in conditions close to the final application. As an experimental tool, a testbed should be flexible to allocate a wide variety of experiments and implementations and should be easy to use.

Moreover, the development of a Cooperating Objects testbed should carefully consider Cooperating Objects characteristics. A detailed description of these characteristics can be found in [1]. Below we briefly summarize some of them:

- **Modularity**: A Cooperating Object is composed of several elements that need to exhibit certain features. Each of the elements contributes to the functionality of the overall Cooperating Object, but the modularization helps to keep the single devices simple and maintainable.
- **Autonomy**: Each element can decide on its own about its involvement in a Cooperating Object. It can also decide about the degree of participation, leaving the possibility to serve multiple Cooperating Objects.
- **Heterogeneity**: (Widely understood) heterogeneity is a crucial point in Cooperating Objects. It may affect all levels from physical characteristics to sensing, computing and communication capabilities.
- **Communication**: Explicit communication is the most frequent in Cooperating Objects. In some cases passive action recognition is an interesting way of information interchange.
- **Common Goal**: The ultimate reason for a Cooperating Object to exist is the common goal it tries to achieve. Although individual Cooperating Object elements might not know the overall goal, they execute a task to achieve it. Thus, each element has detailed knowledge only about its area of responsibility, but limited information about the whole Cooperating Object. It is the cooperation among the elements what makes it possible to achieve the overall goal.

- **Cooperation**: In Cooperating Objects cooperation is always intentional and driven by a goal. The participation of all elements in a Cooperating Object is needed to achieve the common goal: a Cooperating Object is more than just the sum of the single elements.

1.3 Book Structure

This book describes a testbed for Cooperating Objects that allows full cooperation and interoperability between heterogenous elements from different technological fields. It was developed in the CONET Network of Excellence (http://www.cooperating-objects.eu) partially funded by the European Commission under grant INFSO-ICT-224053. The main objective of the CONET Integrated Testbed is to become a benchmark to facilitate comparison and assessment of Cooperating Objects techniques and algorithms from academic and industrial communities.

The CONET Cooperating Objects Integrated Testbed includes a fleet of mobile robots, a WSN with static and mobile nodes and a camera network, the most frequently used elements in Cooperating Objects techniques and experiments. It comprises sensors with high level of heterogeneity including static and mobile cameras, laser range finders, GPS receivers, accelerometers, temperature sensors, light intensity sensors, microphones, among others. The testbed has been developed with an open and modular architecture, which allows equanimity among these elements independently of their sensing, computing or communication capabilities. It employs standard tools and abstract interfaces, increasing its usability and the reuse of code and making the addition of new hardware and software components fairly straightforward. Its architecture enables experiments with different levels of decentralization, including fully distributed and centralized experiments.

The CONET Cooperating Objects Integrated Testbed includes a set of tools to improve usability and support experiment development, execution and analysis. It uses a set of basic functionalities in order to release the users from programming the modules that may be unimportant in his particular experiment, allowing them to concentrate on the techniques to be tested. Installed in a room of $500\,m^2$ at the basement of the School of Engineering of Seville (Spain), and in operation since 2010, the testbed can be fully operated remotely through the Internet in a secure way using a Graphic User Interface.

Besides the introductory and concluding chapters, this book is structured in the following four chapters. Chapter 2 presents the state of the art in testbeds in Cooperating Objects technological fields. Instead of exhaustively describing all testbeds, its objective is to survey the different approaches and schemes focusing on interoperability among elements from different technological fields. Special interest is devoted to existing tendencies.

Two main conclusions can be derived from this survey: (1) there is a clear trend in all individual Cooperating Objects technological fields to increase the level of heterogeneity and interoperability in testbeds including elements with higher level of mobility and ubiquity; (2) despite the large number of testbeds developed, most of them concentrate on cooperation between elements of the same technological field and few of them allow interoperability between elements from different Cooperating Objects fields.

Chapter 3 presents the architecture of the CONET Integrated Testbed including its main requirements and its main hardware components.

Chapter 4 is devoted to the tools developed to improve usability of the CONET Integrated Testbed including basic functionalities straightforward usable by testbed users, tools for remote access such as the Graphical User Interface, testbed radio characterization models and simulators to facilitate experiment development.

Finally, Chap. 5 presents some of the experiments carried out in the testbed in order to illustrate its experimentation capabilities. They include experiments dealing with cooperation among elements from the same technological field, such as multi-robot experiments and WSN experiments, as well as experiments dealing with cooperation between elements from different Cooperating Objects technological fields.

References

1. Marrón PJ, Minder D, Karnouskos S (2012) The emerging domain of cooperating objects: definition and concepts. Springer, New York, http://www.springer.com/engineering/signals/book/978-3-642-28468-7
2. Pedro J, Marrón DM, Stamatis K, Ollero A (2011) The emerging domain of cooperating objects. Springer, New York
3. Marrón PJ, Karnouskos S, Minder D and the CONET Consortium, editor (2009). Research Roadmap on Cooperating Objects. Office for Official Publications of the European Communities, Luxembourg. ISBN 978-92-79-12046-6.

Chapter 2
Testbeds for Cooperating Objects

2.1 Introduction

In the last years the interest in Cooperating Objects technologies has originated a growing demand of tools for testing and validating algorithms and methods. Lately, the development of testbeds for technologies such as multi-robot systems and ubiquitous and pervasive devices has been intensified both in number and variety.

This chapter briefly presents the existing testbeds for Cooperating Objects. Cooperation between heterogenous devices and technologies is a key issue in the Cooperating Objects domain. Thus, we classify testbeds attending to the level of integration between different technological fields. Other features such as heterogeneity, flexibility, modularity and ease of use are also analyzed. The selection of the software architecture and middleware is critical for the modularity, flexibility and interoperability of the testbed. This chapter also reviews the main existing middleware for integration of Cooperating Objects.

This chapter is structured as follows. Section 2.2 briefly analyzes Cooperating Objects testbeds attending to their level of integration between elements from different technological fields. Section 2.3 reviews the main existing middleware for Cooperating Objects. Existing Cooperating Objects testbeds with no integration are briefly summarized in Sect. 2.4. Testbeds with low level of integration are analyzed in Sect. 2.5. Testbeds for Cooperating Objects with high level of integration are reviewed in Sect. 2.6. Finally, Sect. 2.7 concludes the chapter.

2.2 Components of a Cooperating Objects Testbed

The testbeds analyzed in this chapter integrate elements from different Cooperating Objects technological fields ranging from robotic systems to ubiquitous systems such as Wireless Sensor Networks (WSN), camera networks and networks for smartphones, PDAs and tablets, among others.

J. R. Martinez-de Dios et al., *A Remote Integrated Testbed for Cooperating Objects*,
SpringerBriefs in Cooperating Objects,
DOI: 10.1007/978-3-319-01372-5_2, © The Author(s) 2014

Wireless Sensor Networks are comprised of a set of embedded nodes with sensing, processing and communication capabilities. WSN nodes can be integrated with a wide variety of sensors such as temperature, humidity, light intensity sensors and accelerometers, GPS receivers and even small cameras such as Cyclops and CMUcam3. WSN radio circuitries also allow to measure the strength of the radio signal of the received messages (RSSI) and other Link Quality Estimators. WSN nodes used in testbeds are frequently static, although mobile nodes may exist if carried by people or mounted on mobile robots. The design of networking and routing algorithms supporting mobility is one important open research area in WSN. Other components such as camera networks, PDAs and smartphones can also be frequently found integrated in these testbeds.

A wide variety of robotic platforms have been integrated in testbeds. Ground robots are usually able to carry heavier payloads, and tend to have larger on-board sensing and processing capabilities. On-board sensors include cameras—from regular color cameras to RGB-D cameras, laser range finders, ultrasound sensors, GPS receivers and Inertial Navigation System units, among others. Aerial robots, although limited in payload, can move in 3D. Quad-rotors, with vertical take-off and landing capabilities, are the most commonly used platforms in testbeds, although blimps, fixed-wing platforms and helicopters are also found mainly in outdoors testbeds. Underwater or surface marine vehicles are rarely used in multi-vehicle testbeds although there are some exceptions.

A rich variety of sensors can be integrated in these aforementioned platforms. They have high degree of heterogeneity. While low cost, low size and low energy constrain the characteristics of WSN sensors, robots can carry and provide mobility to sensors with higher performance. Testbed communications typically include separate wireless networks for the multi-robot system and for the WSN. They differ greatly in range, bandwidth, quality of service and energy consumption. WSN networks were designed for low-rate and low-range communications whereas Wi-Fi networks, typically used by robotic systems, can provide significantly higher rates at greater distances.

A testbed for Cooperating Objects should share design space with Cooperating Objects themselves. We have already identified several Cooperating Objects characteristics, which are depicted in more detail in [1]. More specifically, when designing Cooperating Objects testbeds we have to deal with issues such as interoperability and level of integration among heterogeneous technologies, flexibility, modularity, extensibility, scalability and usability. Refer to [1] for detailed definitions.

A high percentage of the testbed interoperability, flexibility, modularity, extensibility and scalability depends on its software architecture and middleware. Next section briefly presents a review of the main existing middleware for Cooperating Objects.

The level of integration between elements from different technological fields is one of the main characteristics of Cooperating Objects testbeds. An ideal testbed should allow high degree of interoperability. This interoperability is not easy due to the high diversity in their sensing, actuation, computational and communication capabilities. Nevertheless, heterogeneity is in many cases the origin of interesting synergies. Thus, according to the level of integration testbeds can be classified in *testbeds with no integration*, *testbeds with partial integration* and *full Cooperating Objects testbeds*.

2.3 Middleware for Cooperating Objects

The main functions of a middleware for Cooperating Objects can be enumerated as follows:

- support communication among heterogeneous components,
- enable interoperability and integration mechanisms among objects,
- offer often-needed system services and functionalities,
- simplify the development process.

A high number of middleware for robotic and ubiquitous systems have been developed. The higher complexity of robotic technologies suggests adopting a robotic middleware extended with capabilities to integrate WSN and other ubiquitous systems. Below is a brief review of the main existing middleware for networked robotics systems.

Robot Operating System [ROS]

ROS[1] is an open-source software framework for robot software development that provides functionalities that one could expect from an operating system. ROS includes functions for hardware abstraction, low-level device control, implementation of commonly-used functionalities, message-passing between processes and package management, among others. ROS is based on a graph architecture where processing takes place in nodes that may receive, post and multiplex sensor, control, state, planning, actuator and other messages. It includes a wide range of functionalities to help development such as simultaneous localization and mapping (SLAM), planning, image processing and perception, simulation and logging, among others. ROS also includes a set of simulators, such as Stage [2] and Gazebo [3]. ROS has been recently accepted as a "*de facto*" standard by the robotics community enabling code reuse and sharing.

Player/Stage

Player/Stage[2] provides a server-client infrastructure with hardware drivers and basic algorithms for mobile robotic [4]. Player abstracts hardware heterogeneity by using standard interfaces to communicate low-level server drivers and high-level client programs. The server drivers interact with the actual sensors and actuators. Various client-side libraries exist in the form of proxy objects for different programming languages (C, C++, Java, Matlab and Python, among others) to access the

---

[1] http://www.ros.org

[2] http://playerstage.sourceforge.net

services provided by the Player platform. Clients can connect to Player servers to access data, send commands or request configuration changes to an existing device.

Its modular architecture makes it flexible to support new hardware. Player is independent of platforms, programming languages and transport protocols. It includes a ready-to-use set of drivers for most commercial platforms and sensors, a set of algorithms for perception and robot navigation and a number GUIs, logging and visualization tools. It is also compatible with Stage and Gazebo.

YARP

YARP is an open-source set of libraries, protocols for development of multi-robot applications [5]. It provides support to the robot control system as a collection of programs communicating in a peer-to-peer way, with a family of connection types. It includes a number of tools and functionalities aiming to keep modules and devices cleanly decoupled in a modular architecture. YARP also supports flexible interfacing with hardware devices. Written in C++, YARP is OS neutral and provides support for common devices used in robotics such as framegrabbers, digital cameras, motor control boards, among others.

Miro

Miro [6] is an object-oriented middleware developed at University of Ulm. It uses the common object request broker architecture (CORBA) standard. This allows inter-process and cross-platform interoperability for distributed robot control, using multi-platform libraries for easy portability. Examples of these libraries are the CORBA based Adaptive Communication Environment (ACE) [7] or the CORBA Notification Services [8].

Miro allows the implementation of multi-tier architectures. Its provides object-oriented interface abstractions for many sensors and actuators. It also provides a number of often-needed services such as robot mapping, self localization, behavior generation, path planning, logging and visualization facilities. The layered architecture and object-oriented approach make Miro significantly flexible and extendible [9].

Orocos

The Orocos 2.0 toolchain [10] is used to create real-time robotics applications using modular, run-time configurable software components. It provides multi-platform support, extensions to other robotics frameworks such as ROS and YARP, code generators to transfer user-defined data between distributed components, run-time and real-time configurable and scriptable components and event data logging and reporting, among others.

Orca

Orca[3] is a component-based branch of Orocos developed at KTH. It is an open-source framework for developing component-based robotic systems that provides the means to define and develop building-blocks that can be pieced together to form from single vehicles to distributed sensor networks [11]. It enables software reuse

---

[3] http://orca-robotics.sourceforge.net

by defining a set of commonly-used interfaces, providing libraries with a high-level application programming interface (API) and maintaining a repository of components. Orca adopts a Component-Based Software Engineering approach without applying any additional architectural constraints. It uses a commercial open-source library for communication and interface definition and provides tools to simplify component development and uses cross-platform development tools.

Microsoft Robotics Developer Studio

Microsoft RDS[4] is a Windows-based environment for academic and commercial developers to help in the development of robotic applications. It includes a lightweight asynchronous service-oriented runtime and a set of visual authoring and simulation tools, as well as templates, tutorials and sample code. It includes a development environment with a wide range of support to ease the development of robot applications. Reference Platform design defines a minimum set of components and associated software services that will allow a user to begin using RDS with relatively little effort.

Carnegie Mellon Robot Navigation Toolkit [CARMEN]

CARMEN[5] is an open-source modular software designed to provide basic navigation primitives including: base and sensor control, logging, obstacle avoidance, localization, path planning, and mapping for mobile robots [12]. CARMEN provides robot hardware support for some platforms and sensors. It also includes a 2D robot/sensor simulator, process monitoring, message logging and playback functionalities. Communications between CARMEN programs are handled using a separate package called IPC. CARMEN runs under Linux. It is written in C and also provides support for Java.

Robot Software Communication Architecture [RSCA]

RSCA is a middleware for networked robots developed by Seoul National University [13]. It provides a standard operating environment and development framework for robot applications and includes mechanisms to ensure Quality of Service. It consists of a Real-Time Operating System compliant with PSE52 in IEEE POSIX.13 and a communication middleware compliant with CORBA and RT-CORBA v.1.1 [14]. RSCA provides an abstraction layer that makes robot applications portable and reusable on different hardware.

Mobile and Autonomous Robotics Integration Environment [MARIE]

MARIE is a flexible distributed system that allows developers to share, reuse, and integrate robotic software programs for rapid application development [15]. It uses the ACE [7] communication libraries as integration infrastructure. There are four functional components: application adapters, application managers, communication adapters and communication managers. Its architecture allows interoperability among elements and re-usability of application components.

---

[4] http://www.microsoft/robotics

[5] http://carmen.sourceforge.net

Robot-Technology (RT) middleware

The RT middleware [16] uses CORBA to build robot software in a modular structure, simplifying this process by combining selected modules. A further feature of this middleware is to allow the distribution of robot resources over a network.

Sensory Data Processing Middleware

The Sensory Data Processing Middleware [17] abstracts services to access sensor information and support service mobile robots. It provides a unified model for different configurations of external sensors on a mobile robot. Developed services can be reused in applications without dealing with individual sensors. Two types of services are implemented to provide obstacle information and to localize the robot position using landmark observations from multiple external sensors.

Universal Plug and Play (UPnP) Robot middleware

The UPnP Robot middleware [18] was developed to offer peer-to-peer network connectivity among PCs, wireless pervasive devices and intelligent appliances. It uses the Universal Plug and Play—UPnP[6] architecture for dynamic software integration and for control of ubiquitous robots. UPnP mechanisms are used to configure robot components and to allow distributed robots to discover and interact with other devices like cameras and sensor networks. This approach provides a simple scheme for building intelligent robots with a wide variety of hardware and software components.

2.4 Cooperating Objects Testbeds with No Integration

A very large number and variety of testbeds have been developed in each of the individual technological fields within the Cooperating Objects domain. In these testbeds cooperation is constrained to elements belonging to the same field. However, the analysis of their approaches, capabilities and tendencies is very interesting to derive requirements for the design of full Cooperating Objects testbeds. We divide them in static and mobile testbeds. It is not the objective of this section to provide an exhaustive description but to give an overall view illustrating different schemes and approaches.

2.4.1 Static Cooperating Objects testbeds

They basically include WSN testbeds, camera networks testbeds and smarphone testbeds. We focus on the first one in which a high number and variety of approaches can be found.

A very high number of WSN testbeds have been developed. Testbeds are the most popular experimental tools in WSN community. Many of them have been designed

---

[6] http://www.upnp.org

as a service to the research community: they are modular, can allocate a very wide range of experiments and are open to the public. Others have been developed to be focused on particular applications or scenarios. The main tendencies in current WSN testbeds include the following:

- federations of testbeds,
- outdoor testbeds in real-world settings,
- integration with robots and other actuation devices.

TWIST [19] at Technische Universität Berlin is a good example of a publicly accessible heterogeneous WSN testbed. It is comprised of 102 TmoteSky nodes and 102 eyesIFX nodes deployed in a $1,500\,m^2$ scenario in three floors of an instrumented office. TWIST allows public remote access through a simple HTML interface. Its software architecture is modular, flexible and highly extensible, being able to hold a wide range of experiments. In fact, its architecture has been used in other testbeds such as WUSTL [20]. Other publicly accessible general purpose WSN testbeds are: DES-Testbed [21] at Frei Universität Berlin; SensorNet [22] in the Intel Research Center at University of California at Berkeley, comprised of 97 Crossbow Mica2 and 51 Mica2Dot nodes; NetEye [23] with 130 TelosB nodes; and the INDRIYA testbed [24] at University of Singapore, see Fig. 2.1.

Besides, many testbeds have been developed to meet particular needs or functionalities, losing generality but gaining efficiency. Networking is a commonly targeted research topic in these functionality-driven WSN testbeds. Some examples are: WiN-TER testbed [25] at Acadia University, NESC testbed [26] and VineLab testbed [27]

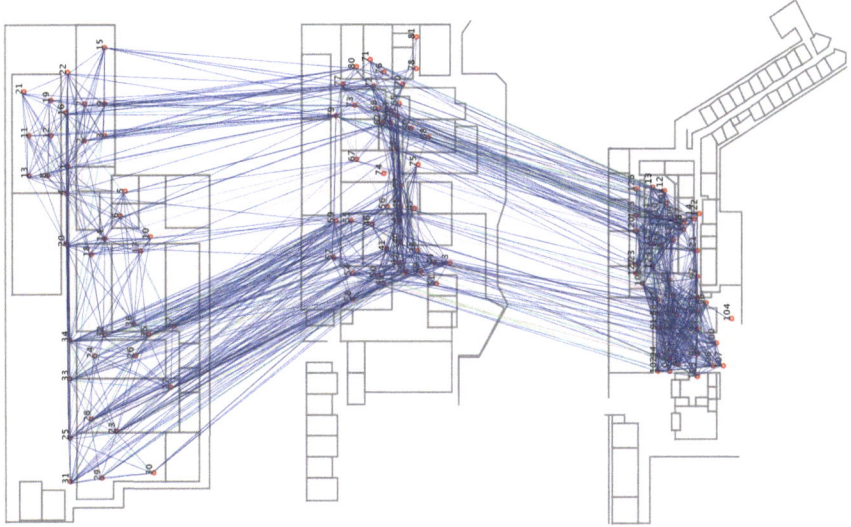

Fig. 2.1 Typical WSN node connectivity in INDRIYA testbed (courtesy National University of Singapore)

at University of Virginia. BANAID [28] targets security and worm-hole attack problems. The Imote2 testbed [29] deals with localization using RSSI from a grid of Crossbow Imote2 nodes. RadiaLE at ISEP [30] is a remotely and publicly accessible outdoors testbed that targets radio management and link quality estimators.

One of the latest tendencies is to group several testbeds under a federation. Various testbeds located at different places can be used transparently as one, benefiting from the variety of capabilities/setting each one provides. Federated testbeds share resources -virtually or physically- and can be accessed with the same API, which allows for instance testing the same experiment in different settings. The number of WSN nodes in federated testbeds ranges between around 500 in KanseiGeni [31] to 1,024 in SensLab.[7] Other examples are XSensor [32], WISEBED [33] and COFEDE.[8] It is interesting to mention the *FP7 Future Internet Research and Experimentation* (FIRE) program [34] in which a number of testbeds with a common architecture are in development. The FIRE architecture has been defined to ensure compatibility among all the FIRE testbeds.

One recent trend is the development of large outdoors testbeds for experimentation in real-world conditions. TeamTrack [35] is a characteristic example. It includes fully-instrumented tablet kits—equipped with GPS, camera and digital compass—and lightweight handheld kits (an HP handheld and Bluetooth GPS unit). TeamTrack is fairly scalable and mobility is provided by people carrying the nodes. CitySense [36] is an open testbed at Harvard University that integrates 100 Linux-based embedded PCs with dual IEEE 802.11a/b/g radios and custom sensor nodes. SmartSantander[9] is an European project in which 20,000 nodes based on XBee-PRO will be deployed in four European cities, see Fig. 2.2.

2.4.2 Mobile Cooperating Objects testbeds

They are basically mobile robot testbeds. A very high number of robot and multi-robot testbeds have been developed along the years. Experimentation in real-world conditions is critical in robotics research due to the difficulties to simulate physical interaction. Most robotic testbeds have been developed specifically to test target functionalities or applications. A number of testbeds capable of allocating a wider range of experiments. The main tendencies in current networked robots testbeds include the following:

- testbeds of aerial robots,
- integration of robots with different mobility capabilities,
- integration with ubiquitous systems.

---

[7] http://www.senslab.info

[8] http://www.cotefe.net

[9] http://www.smartsantander.eu

Fig. 2.2 General view of node deployment in the city of Santander, Spain (courtesy Telefonica I+D)

In robotics the need to test the methods in real conditions has motivated the development of a high number of robot testbeds designed to address specific functionalities or applications. Only some examples are to be cited. The Micro Autonomous RoverS (MARS) testbed, used in [37], at Stanford University, comprised of six custom made rovers, was developed for research in human–robot interaction, motion planning and object tracking. The CALTECH Multi-vehicle Wireless Testbed [38] includes 8 custom hovercrafts. CALTECH testbed targeted experiments in multi-robot coordination, networked control and real-time networking. Other functionality driven robot testbeds are: UTIAS testbed at University of Toronto, used to test the techniques proposed in [39]; the testbed used to validate the method in [40], at the Massachusetts Institute of Technology (MIT), which targets multi-robot exploration; the testbed used in [41] at Kansas State University, which focuses on search and rescue applications; [42] at Columbia University, which targets network connectivity and data collection; and the testbed used to validate [43] at University of California-Los Angeles (UCLA), which focuses on boundary tracking and estimation.

The need to compare and evaluate different methods in the same conditions motivated the development of general purpose testbeds. General purpose testbeds are generally designed to be used by a broader public than functionality-driven testbeds. This involved efforts to have architectures with higher modularity, flexibility and extensibility. It also involved higher interest in code reuse, openness approach and the use of common languages and interfaces. [44] describes a testbed in the GRASP Laboratory at the University of Pennsylvania based on Player/Stage [2]. Its architecture allows good scalability and easy operation even with large deployments of mobile robots. The COMET testbed [45] integrates 10 robots under a flexible and extensible architecture. The Autonomous Mobile Robotics Systems (AMRS) testbed [46] at University of Louisville also uses Player/Stage as supporting middleware.

The AMRS testbed is comprised of customized platforms while the GRASP testbed includes SCARAB robotic platforms.

The generalization of robot testbeds as tools for methods evaluation involved a higher need for remote access and used-friendly operation. The Mint-m testbed [47], developed at the Stony Brook University, comprised of 12 iRobot Roomba platforms controlled through a central processor that uses overhead cameras to compute robot poses. The testbed uses an ad-hoc software architecture with high flexibility and reconfigurability. It can be remotely operated and has tools for experiment visualization and logging. The testbed also offers tools for experiment simulation. Similarly, HoTDeC [48] at University of Illinois at Urbana-Champaign provides a GUI for remote visualization and monitoring and allows remote configuration and high level robot control.

The aforementioned testbeds integrate only ground robots. In the last years a number of testbeds using Unmanned Aerial Vehicles (UAV) and Autonomous Underwater Vehicles (AUV) have been developed. Below are some examples: [49] at MIT Space Systems and Artificial Intelligence Laboratories uses 3 custom flying crafts; [50] at University of Southern California (USC) integrates customized UAVs based on helicopters; the GRASP Multiple MAV testbed [51] includes *Ascending Technologies* Hummingbird quad-rotors[10]; and the CATEC[11] testbed is comprised of 10 quadrotors, see Fig. 2.3.

Some testbeds integrating aerial and ground robots have been developed. An example is RAVEN [52] at MIT, which integrates Draganflyer V Ti Pro quad-rotors, a foam R/C aircraft and modified DuraTrax RC trucks. MAGICC Lab. [53] developed

Fig. 2.3 General overview of CATEC testbed (courtesy FADA-CATEC)

[10] http://www.asctec.de

[11] http://www.catec.com.es

a testbed with 5 custom ground robots, one omnidirectional robot and 4 fixed-wing UAVs. The testbed used in [54] at MIT, equipped with 2 custom remotely operated underwater vehicles, is an example of an underwater multi-robot testbed.

2.5 Cooperating Objects Testbeds with Low Integration

This type of testbeds allows some interoperation between heterogenous Cooperating Objects technologies. They are often designed so that one of the technologies—primary—has control over the other. Therefore, the balance of experiments that can be held is biased either towards ubiquitous systems or towards robotics and only part of the features of the secondary technology are exploited. In some cases the testbed focus is on WSN and robots are essentially used to provide mobility to some nodes. Others are focused on robotics: they are robot testbeds enhanced with a WSN, which is essentially used as a distributed sensor or as communication backbone. Again, the objective of this section is to provide an overview of the different approaches and trends rather than an exhaustive enumeration.

The Hybrid sensor network testbed at Deakin University was developed to test localization methods in delay-tolerant sensor networks. It is comprised of static Mica2 WSN nodes and Lego robots. The robots locate the static nodes using RSSI while they move following paths. Each robot carries one Stargate board that receives RSSI readings from the static WSN and executes the localization method. There is no explicit communication between each robot and its Stargate board: robots are used to carry the boards along predefined paths.

Mobile Emulab [55] is a general purpose WSN testbed that uses robots to provide mobility to nodes. Developed at University of Utah, it is comprised of 5 *Acroname* Garcia robots and a static WSN with 25 Mica2 nodes. Each robot carries one WSN node but there is no communication between them.

Mobile Emulab includes simple robot obstacle avoidance and path-planning methods but the testbed focus and main objective are research in WSN. The testbed uses an open and modular architecture and includes a GUI for remote operation and provides high degree of interaction with the user and can show live maps and images of the experiment. It was open for public use until 2008.

The iMouse testbed [56] targets cooperative surveillance integrating a WSN and a fleet of ground robots. The WSN operates as a distributed sensor that triggers events to be monitored by the robot fleet. Coordination among systems is achieved through a central server that receives events from the WSN and assigns robots to event locations: there is no direct communication between robots and WSN. The robots are based on *Lego* Mindstorm platforms enhanced with a Stargate board and a webcam, whereas its WSN is comprised of 17 MicaZ nodes. The iMouse has a centralized architecture and uses an ad-hoc software architecture.

Sensei UU [57] targets localization experiments. It is comprised of *Lego* NXT based robots to enhance a static WSN with repeatable mobility. Each robot includes a TelosB node and one smartphone, which acts as the main robot processor. The testbed uses two networks: one based on IEEE 802.11b/g (Smartphones) and one based on IEEE 802.15.4 (WSN nodes). Robot motion is controlled from the central site manager, which transmits commands to the robot through the smartphone. The robot platform reports the estimated traveled distance to the smartphone, which uses it to determine the robot position. The WSN node reports the received RSSI to the smartphone using simple unidirectional messages. Robot motion is constrained to fixed paths. Markers on the tracks are used for robot localization.

The Kansei testbed [58] is comprised of one static WSN with 210 nodes and *Acroname* mobile robots. Each static node includes one Extreme Scale Mote (XSM) and one Stargate board. Robots are equipped with one XSM that communicates with the static WSN. Robots and WSN are integrated by the testbed Director, a centralized software platform that enables uniform, integrated experimentation. The testbed has a modular, extensible and scalable architecture. It also includes software components for job management in complex multi-tier experiments. It can be accessed remotely and includes a hybrid hardware–software simulation engine to facilitate experimentation.

Work [59] describes a testbed for multi-robot coordination experiments comprised of heterogeneous ground robots including micro vehicles enhanced with custom sensing and wireless communication devices. Robotic platforms are connected to WSN nodes through a serial port. These nodes provide computing and communication capabilities and send direct commands to the robotic platforms. The testbed is equipped with overhead cameras to compute robots ground-truth poses. Also, an optional data monitor can be used for experiment logging and visualization. Its uses an ad-hoc software architecture with good flexibility and allows distributed and centralized experiments.

The Explorebots testbed [60] focuses on experimentation of protocols for mobile multi-hop networks. It uses robots based on customized Rogue ATV platforms, each carrying a Mica2 node as communication module. The robots are equipped with webcams, electronic compasses and range sensors. Each robot has a WSN node that provides commands to and obtains information from it using a simple protocol. The testbed uses a closed architecture with low flexibility and extensibility. A GUI allows on-site visualization of images and data.

Robomote [61] is a general purpose testbed comprised of custom micro robotic platforms—Robomote—equipped with simple compass and proximity sensors. A Mica2 WSN node acts as the master of each mobile node. The Robomote processor is in charge of the communication with the sensors and actuators and with the Mica2, which implements higher level sensing and mobility control functions. It uses a modular and open distributed software architecture based on TinyOS.

2.6 Cooperating Objects Testbeds with High Integration

They have the capacity to hold Cooperating Objects balanced experiments in which different heterogeneous devices from different technological fields work as peers. Few existing testbeds comprise the necessary hardware and software infrastructure to belong to this category, and the majority of them are often focused on particular functionalities, with the corresponding loss of generality. Different designs of a general Cooperating Objects testbed have been proposed in a number of works. However, in few of them the testbed is implemented and made open to the community. UBIROBOT[12] comprises WSN, PDAs, smartphones and mobile robots equipped with a variety of sensors. Apart from a general design we could not find further scientific publications, details or information on its current availability. In some other cases, the approach and capabilities are promising but the testbed was developed specifically for experimentation of some functionalities, which limits its impact in the community. That is the case of the SWARM testbed [62], which focused on swarm cooperation between a fleet of robots and a static WSN, considered also as a swarm.

The testbed for networked Physically Embedded Intelligent Systems (PEIS) targets methods for the use of ubiquitous systems in every day live [63]. The PEIS-home testbed, at the AASS Mobile Robotics Laboratory (University of Örebro), is a small apartment equipped with a fleet of robots Magellan Pro robot and Roomba, among others, with communication infrastructure, ubiquitous computing devices, a camera network, automatic appliances and embedded sensors, see Fig. 2.4. It also includes WSN and RFID technologies. The PEIS architecture is centralized modular, flexible and highly extensible and scalable.

ISROBOTNET [64] was developed within the framework of the URUS project on ubiquitous robotics in urban settings. Thus, its design and architecture aim to meet the project needs. The testbed includes a camera network with ten webcams and 5 robots (4 Pioneer 3AT and one ATRV-Jr) equipped with self-localization and obstacle avoidance sensors. WSN nodes are used as distributed sensors for localization and tracking. The testbed uses a modular and flexible service-oriented middleware based on open-source software to simplify the integration of subsystems developed by different users and to enable easy transition from simulation to experimentation.

Fig. 2.4 Pictures from the PEIS testbed (courtesy University of Örebro)

---

[12] http://ubirobot.ucd.ie

It includes software modules with image processing algorithms for human activity detection, people tracking using static and mobile cameras, among others.

2.7 Conclusions

The interest in Cooperating Objects technologies has motivated the development of a high number and variety of experimental testbeds. Despite the number and variety of existing testbeds, most of them focus on cooperation among elements from the same technological field. Very few of them enable full integration and interoperability between elements from different Cooperating Objects fields and, those which do, are focused on specific applications or functionalities. This lack of suitable testbeds for integrated Cooperating Objects research has been pointed out as one major drawback for the development of the Cooperating Objects domain. The CONET Cooperating Objects Integrated Testbed was designed to fill in this gap.

References

1. Marrón PJ, Minder D, Karnouskos S (2012) The emerging domain of cooperating objects: definition and concepts, Springer. http://www.springer.com/engineering/signals/book/978-3-642-28468-7
2. Gerkey BP, Vaughan RT, Howard A (2003) The player/stage project: tools for multi-robot and distributed sensor systems. In: Proceedings of the international conference on advanced robotics, pp 317–323
3. Koenig N, Howard A (2004) Design and use paradigms for gazebo, an open-source multi-robot simulator. In: IEEE/RSJ international conference on intelligent robots and systems, pp 2149–2154
4. Gerkey B, Vaughan R, Stoy K, Howard A, Sukhatme G, Mataric M (2001) Most valuable player: a robot device server for distributed control. In: Proceedings of the IEEE/RSJ international conference on intelligent robots and systems, vol 3. pp 1226–1231
5. Metta G, Fitzpatrick P, Natale L (2006) Yarp: yet another robot platform. Int J Adv Robot Syst 3:43–48
6. Utz H, Sablatnog S, Enderle S, Kraetzschmar G (2002) Miro-middleware for mobile robot applications. IEEE Trans Robot Autom 18(4):493–497
7. Schmidt DC (1994) Ace: an object-oriented framework for developing distributed applications. In: Proceedings of the 6th USENIX C++ technical conference
8. Harrison TH, Levine DL, Schmidt DC (1997) The design and performance of a real-time corba event service. SIGPLAN Not 32:184–200
9. Mohamed N, amd Imad Jawhar JAJ (2009) A review of middleware for networked robots. Int J Comput Sci Netw Secure IJCSNS 9(5):139–148
10. Bruyninckx H (2001) Open robot control software: the orocos project. In: Proceedings of the IEEE international conference on robotics and automation, ICRA2001, vol 3. pp 2523–2528
11. Brooks A, Kaupp T, Makarenko A, Williams S, Oreback A (2005) Towards component-based robotics. In: Proceedings of the IEEE/RSJ international conference on intelligent robots and systems, IROS2005, pp 163–168
12. Montemerlo M, Roy N, Thrun S (2003) Perspectives on standardization in mobile robot programming: the carnegie mellon navigation (carmen) toolkit. In: Intelligent robots and systems,

2003. (IROS 2003), Proceedings of the 2003 IEEE/RSJ international conference on, vol 3. pp 2436–2441. DOI:10.1109/IROS.2003.124923

13. Yoo J, Kim S, Hong S (2006) The robot software communications architecture (rsca): Qos-aware middleware for networked service robots. In: Joint international conference SICE-ICASE, pp 330–335

14. Schmidt D, Gokhale A, Harrison T, Parulkar G (1997) A high-performance end system architecture for real-time corba. IEEE Commun Mag 35(2):72–77

15. Cote C, Brosseau Y, Letourneau D, Raievsky C, Michaud F (2006) Robotic software integration using marie. Int J Adv Robot Syst 3(1):55–60

16. Ando N, Suehiro T, Kitagaki K, Kotoku T, Yoon WK (2005) Rt-middleware: distributed component middleware for rt (robot technology). In: Proceedings of the IEEE/RSJ international conference on intelligent robots and systems IROS2005), pp 3933–3938

17. Takeuchi E, Tsubouchi T (2006) Sensory data processing middlewares for service mobile robot applications. In: Joint international conference SICE-ICASE, pp 1201–1206

18. Ahn SC, Lee JW, Lim KW, Ko H, Kwon YM, Kim HG (2006) Requirements to upnp for robot middleware. In: Proceedings of the IEEE/RSJ international conference on intelligent robots and systems, IROS2006, pp 4716–4721

19. Handziski V, Köpke A, Willig A, Wolisz A (2006) Twist: a scalable and reconfigurable testbed for wireless indoor experiments with sensor networks. In: Proceedings of the 2nd ACM international workshop on multi-hop ad hoc networks: from theory to reality, New York, pp 63–70

20. Sha M, WUSTL wireless sensor network testbed. http://wsn.cse.wustl.edu/index.php/The_WUSTL_Wireless_Sensor_Network_Testbed Accessed in June 2013

21. Günes M, Blywis B, Juraschek F (2008) Concept and design of the hybrid distributed embedded systems testbed. Technical report TR-B-08-10, Freie Universität Berlin

22. Chun BN, Buonadonna P, AuYoung A, Ng C, Parkes DC, Shneidman J, Snoeren AC, Vahdat A (2005) Mirage: a microeconomic resource allocation system for sensornet testbeds. In: Proceedings of the 2nd IEEE workshop on embedded networked sensors, pp 19–28

23. Sakamuri D, Zhang H (2009) Elements of sensornet testbed design. World Scientific Publishing Co, USA, pp 1–36 (chapter 35)

24. Doddavenkatappa M, Chan MC, Ananda A (2012) Indriya: a low-cost, 3d wireless sensor network testbed. In: Testbeds and research infrastructure. Development of networks and communities, Lecture notes of the institute for computer sciences, social informatics and telecommunications engineering, vol 60. pp 302–316

25. Murillo MJ, Slipp JA (2009) Application of winter industrial testbed to the analysis of closed-loop control systems in wireless sensor networks. In: Proceeding of the 8th ACM/IEEE International Conference on Information Processing in Sensor Networks

26. Kejie C, Xiang G, Xianying H, Jiming C, Youxian S (2008) Design and develop of the wsn testbed: Nesc-testbed. In: China science paper online

27. Whitehouse K, The vinelab wireless testbed. http://www.cs.virginia.edu/~whitehouse/research/testbed/ Accessed in June 2013

28. Alzaid H, Abanmi S (2009) A wireless sensor networks test-bed for the wormhole attack. Int J Digit Content Tech Appl 3(3):19–27

29. Poovendran R, Imote2 sensor network testbed. http://www.ee.washington.edu/research/nsl/Imote/ Accessed in June 2013

30. Baccour N, Koubâa A, Jamâa MB, do Rosário D, Youssef H, Alves M, Becker LB, (2011) Radiale: a framework for designing and assessing link quality estimators in wireless sensor networks. Ad Hoc Netw 9(7):1165–1185

31. Sridharan M, Zeng W, Leal W, Ju X, Ramnath R, Zhang H, Arora A (2010) Kanseigenie: software infrastructure for resource management and programmability of wireless sensor network fabrics. Next generation internet architectures and protocols, Cambridge University Press. http://ebooks.cambridge.org/chapter.jsf?bid=CBO9780511920950&cid=CBO9780511920950A115

32. Kanzaki A, Hara T, Ishi Y, Wakamiya N, Shimojo S (2009) X-sensor: a sensor network testbed integrating multiple networks. In: International conference on complex, intelligent and software intensive systems, pp 1082–1087

33. Chatzigiannakis I, Fischer S, Koninis C, Mylonas G, Pfisterer D (2010) Wisebed: an open large-scale wireless sensor network testbed. In: Sensor applications, experimentation, and logistics, lecture notes of the institute for computer sciences, social informatics and telecommunication Engineering, vol 29. Springer, Berlin, pp 68–87

34. EC, FP7 Future internet research and experimentation (FIRE). http://www.ict-fire.eu/home/fire-projects.html Accessed in June 2013

35. Hemmes J, Thain D, Poellabauer C, Moretti C, Snowberger P, McNutt B (2007) Lessons learned building teamtrak: An urban/outdoor mobile testbed. In: Proceedings of the international conference on wireless algorithms, systems and applications, WASA2007, pp 219–224

36. Murty R, Mainland G, Rose I, Chowdhury A, Gosain A, Bers J, Welsh M (2008) Citysense: an urban-scale wireless sensor network and testbed. In: Proceedings of the IEEE conference on technologies for homeland security, pp 583–588

37. Clark CM, Frew EW, Jones HL, Rock SM (2003) An integrated system for command and control of cooperative robotic systems. In: Proceedings of the 11th international conference on, advanced robotics, pp 459–464

38. Cremean L, Dunbar W, van Gogh D, Hickey J, Klavins E, Meltzer J, Murray R (2002) The caltech multi-vehicle wireless testbed. In: Proceedings of the 41st IEEE conference on decision and control, vol 1. pp 86–88

39. Marshall JA, Fung T, Broucke ME, DEleuterio GM, Francis BA, (2006) Experiments in multirobot coordination. Robot Auton Syst 54(3):265–275

40. Williams BC, Kim P, Hofbaur M, How J, Kennell J, Loy J, Ragno R, Stedl J, Walcott A (2001) Model-based reactive programming of cooperative vehicles for mars exploration. In: Proceedings of the international symposium on artificial intelligence, robotics and automation in space

41. Deloach SA, Matson ET, Li Y (2003) Exploiting agent oriented software engineering in cooperative robotics search and rescue. Int J Pattern Recognit Artif Intell 17(5):817–835

42. Reich J, Misra V, Rubenstein D (2008) Roomba madnet: a mobile ad-hoc delay tolerant network testbed. SIGMOBILE Mob Comput Commun Rev 12:68–70

43. Joshi A, Ashley T, Huang Y, Bertozzi A (2009) Experimental validation of cooperative environmental boundary tracking with on-board sensors. In: Proceedings of the American control conference, ACC2009, 2630–2635

44. Michael N, Fink J, Kumar V (2008) Experimental testbed for large multirobot teams. IEEE Robot Autom Mag 15(1):53–61

45. Cruz D, McClintock J, Perteet B, Orqueda O, Cao Y, Fierro R (2007) Decentralized cooperative control—a multivehicle platform for research in networked embedded systems. IEEE Control Syst 27(3):58–78

46. Riggs T, Inanc T, Zhang W (2010) An autonomous mobile robotics testbed: construction, validation, and experiments. IEEE Trans Control Syst Technol 18(3):757–766

47. De P, Raniwala A, Krishnan R, Tatavarthi K, Modi J, Syed NA, Sharma S, Chiueh Tc (2006) Mint-m: an autonomous mobile wireless experimentation platform. In: Proceedings of the 4th international conference on mobile systems, applications and services, MobiSys2006, ACM, New York, USA, pp 124–137

48. Stubbs A, Vladimerou V, Fulford A, King D, Strick J, Dullerud G (2006) Multivehicle systems control over networks: a hovercraft testbed for networked and decentralized control. IEEE Control Syst 26(3):56–69

49. Miller D, Seanz-Otero A, Wertz J, Chen A, Berkowski G, Brodel C, Carlson S, Carpenter D, Chen S (2000) Spheres-a testbed for long duration satellite formation flying in micro-gravity conditions. In: Proceedings of the AAS/AIAA space flight mechanics meeting, pp 167–179

50. Naffin D, Sukhatme G (2002) A test bed for autonomous formation flying. In: Institute for robotics and intelligent systems, Technical, Report IRIS-02-412

51. Michael N, Mellinger D, Lindsey Q, Kumar V (2010) The grasp multiple micro-uav testbed. IEEE Robot Autom Mag 17(3):56–65
52. How J, Bethke B, Frank A, Dale D, Vian J (2008) Real-time indoor autonomous vehicle test environment. IEEE Control Syst 28(2):51–64
53. McLain T, Beard R (2004) Unmanned air vehicle testbed for cooperative control experiments. In: Proceedings of the American control conference, ACC2004, vol 6. pp 5327–5331
54. Brundage H, Cooney L, Huo E, Lichter H, Oyebode O, Sinha P, Stanway M, Stefanov-Wagner T, Stiehl K, Walker D (2006) Design of an rov to compete in the 5th annual mate rov competition and beyond. In: Proceedings of the OCEANS 2006, pp 1–5
55. Johnson D, Stack T, Fish R, Flickinger DM, Stoller L, Ricci R, Lepreau J (2006) Mobile emulab: a robotic wireless and sensor network testbed. In: Proceedings of the 25th IEEE international conference on, computer communications, pp 1–12
56. Tseng YC, Wang YC, Cheng KY, Hsieh YY (2007) Imouse: an integrated mobile surveillance and wireless sensor system. Computer 40(6):60–66
57. Rensfelt O, Hermans F, Gunningberg P, Larzon L (2010) Repeatable experiments with mobile nodes in a relocatable wsn testbed. In: Proceedings of the 6th IEEE international conference on distributed computing in sensor systems, Workshop, pp 1–6
58. Arora A, Ertin E, Ramnath R, Nesterenko M, Leal W (2006) Kansei: a high-fidelity sensing testbed. IEEE Internet Comput 10(2):35–47
59. Leung K, Hsieh C, Huang Y, Joshi A, Voroninski V, Bertozzi A (2007) A second generation micro-vehicle testbed for cooperative control and sensing strategies. In: Proceedings of the American control conference, ACC2007, pp 1900–1907
60. Dahlberg TA, Nasipuri A, Taylor C (2005) Explorebots: a mobile network experimentation testbed. In: Proceedings of the ACM SIGCOMM workshop on experimental approaches to wireless network design and analysis, ACM, New York, USA, pp 76–81
61. Dantu K, Rahimi M, Shah H, Babel S, Dhariwal A, Sukhatme G (2005) Robomote: enabling mobility in sensor networks. In: Proceedings of the 4th international symposium on information processing in sensor networks, IPSN 2005, pp 404–409
62. Li W, Shen W (2011) Swarm behavior control of mobile multi-robots with wireless sensor networks. J Netw Comput Appl 34(4):1398–1407, advanced topics in cloud computing
63. Saffiotti A, Broxvall M (2005) Peis ecologies: ambient intelligence meets autonomous robotics. In: Proceedings of the joint conference on smart objects and ambient intelligence: innovative context-aware services: usages and technologies, ACM, New York, USA, pp 277–281
64. Barbosa M, Bernardino A, Figueira D, Gaspar J, Goncalves N, Lima P, Moreno P, Pahliani A, Santos-Victor J, Spaan M, Sequeira J (2009) Isrobotnet: a testbed for sensor and robot network systems. In: Proceedings of the IEEE/RSJ international conference on intelligent robots and systems IROS2009, pp 2827–2833

Chapter 3
CONET Integrated Testbed Architecture

3.1 Introduction

The objective of this chapter is to present the architecture of the CONET Integrated Testbed for Cooperating Objects. Interoperability, flexibility and usability were the main requirements in the development of the CONET Integrated Testbed. Its software architecture was developed to enable the interaction of heterogeneous devices with wide differences in sensing, computing and communication capabilities. On the other hand, the testbed was devised to fulfil a generalist approach, i.e. to support a wide range of experiments in as many applications as possible. The solution adopted was to use an integrating layer through which all modules intercommunicate using standardized interfaces.

The CONET Cooperating Objects Integrated Testbed was developed to integrate the most frequently used platforms (static and mobile), communication infrastructure and sensors used in Cooperating Objects experiments. A rich variety of sensors were integrated in static and mobile platforms including cameras, laser range finders, ultrasound sensors, GPS receivers, accelerometers, temperature sensors, microphones, among others.

This chapter starts summarizing a list of requirements and specifications in Sect. 3.2. Section 3.3 describes the proposed software architecture, developed to meet these requirements, devoting special attention to interoperability among heterogeneous devices. Section 3.4 describes the main platforms, devices and sensors selected. Section 3.5 concludes the chapter.

3.2 Specifications

This section infers a set of specifications for the CONET Integrated Testbed. Chapter 2 surveyed the existing Cooperating Objects testbeds and middleware. Its conclusion highlights that the increase in the level of integration and interoperability

J. R. Martinez-de Dios et al., *A Remote Integrated Testbed for Cooperating Objects*,
SpringerBriefs in Cooperating Objects,
DOI: 10.1007/978-3-319-01372-5_3, © The Author(s) 2014

is maybe the clearest tendency among Cooperating Objects testbeds. There is a natural tendency towards the integration of complementary heterogeneous components with a variety of mobility, computing and sensing capabilities. There is a clear tendency in WSN testbeds to increase their level of integration with mobile devices. Robot testbeds also tend to increase their heterogeneity integrating ground and aerial robots as well as ubiquitous systems such as WSN and smartphones. These tendencies involve an increasing effort to improve testbed flexibility and extensibility, by adopting common tools and interfaces and by creating testbed federations. These conclusions have also been pointed out by other surveys and analyses in multi-robot [1], WSN [2] and Cooperating Object domains [3].

The analysys of existing testbeds and middleware summarized in Chapter 2 together with Cooperation Objects properties described in Chapter 1 have been crucial to specify the requirements for the CONET Integrated Testbed. The testbed requirements are also supported by questionnaires to potential users drawn from the CONET Network of Excellence (http://www.cooperating-objects.eu) partners as representatives of the European Cooperating Objects communities from academy and industry. The questionnaires were also responded by CONET Research Clusters, groups of CONET members researching in specific topics such as mobility or deployment of Cooperating Objects, among others.

A survey including on-line questionnaires among these potential users was carried out to identify needs, requirements and specifications. The questionnaire collected information on the main architectural, hardware and usability requirements that the Cooperating Objects testbed should have in order to fulfil their experimentation needs. The questions also tried to determine the type of experiments that potential users would like to perform in the CONET Integrated Testbed.

A total of 84 answers were collected and processed. It is worth highlighting that there were no significant differences between the responses from CONET partners and those from CONET Research Clusters, reflecting the fact that there is common agreement in the requirements from different Cooperating Objects communities despite the heterogeneity of their research interests.

3.2.1 Main Intended Experiments

According to the results of the questionnaire the most frequent methods to be experimented in the Cooperating Objects testbed can be summarized as follows:

- localization techniques,
- data muling strategies with multiple robots,
- optimal robot motion control to improve perception,
- analysis of WSN protocols at different layers,
- analysis of QoS and radio link quality estimation,
- protocol performance in robot-WSN applications,
- evaluation of the impact of external interference on WSN.

Below are some exemplary experiments of these methods.

Data mule and communication optimization strategies

The experiment consists of a swarm of robots—each has one attached WSN node—moving in the scenario in a cooperative way collecting data from static WSN nodes that are disconnected from the WSN base. The robots should move to accomplish navigation goals while collecting data from WSN nodes. The objective is to analyze different strategies and methods to use the robots movement in order to combine data collection and robot navigation.

Experiments on cooperative localization and perception

The objective is to localize and track a mobile target using ubiquitous devices while trying to optimize energy consumption. The objective is to determine the minimal set of ubiquitous devices that should be active to track a mobile target. Tracking of targets under uncertainty using a fleet of networked robots is also considered. It could also include methods for simultaneous localization and mapping (SLAM).

Default resource management and adaptation scenario

A person carrying a Body Area Network (BAN) moves and passes by several sources of interference. The objective is to determine which channels are the most suitable for communication and robust to interference. In this experiment repeatability is important. The person carrying the BAN might be emulated by one or several robots carrying one or more sensor nodes that form the BAN.

The most frequent experiments resulting from the questionnaires included in roughly the same percentage the use of static networked devices exclusively and the use of static devices in combination with mobile robots. Around 30 % of the users expressed their need of using heterogeneous WSN nodes in their experiments. 65 % of them required temperature sensors and RSSI in static devices. A smaller percentage requested some other WSN sensors, such as light, accelerometers, etc. Odometry and global position were the top requested measurements for mobile devices. Also, 40 % of the users needed to use camera images and laser readings. Finally, some users requested the possibility to extend the robot fleet with heterogeneous indoor and outdoor robots (aerial robots).

The main requirements regarding operation were flexibility and usability. The questionnaires did not show any preferences in robotic operating systems or middleware, although the top requested operating systems in WSN were Contiki and TinyOS. Most potential users found very interesting the possibility of using simple commands such as experiment start/stop and logging start/stop. Finally, some of the intended experiments required specific setting such as obstacles, sensor nodes located at specific (user defined) locations, etc. Thus, flexibility in the configuration of the scenario was also identified as a requirement.

The questionnaire responses regarding experiment visualization and monitoring tools showed that most users needed to visualize node positions and sensor values, robot odometry and position and general experiment views. The need of providing the Graphic User Interface (GUI) with capabilities to visualize robot laser readings and images was also identified. The survey revealed that not only graphical data but also plain text messages were considered necessary.

3.2.2 Main Requirements

Below is a summary of the main requirements of the CONET Cooperating Objects Integrated Testbed. The main architecture requirements identified were:

- **interoperability between heterogeneous technologies**, with different sensing, computing and communication capabilities,
- **openness**, compatibility with existing devices, standards and operating systems,
- **flexibility**, allowing centralized, decentralized and hybrid experiments,
- **scalability and extensibility**, being adaptable to new hardware and software components,
- **reliability and robustness**,
- **timeliness**, avoiding delays in communications and allowing real-time operation.

    The main hardware requirements were:

- **availability of COTS static node sensors**, including temperature sensor and light sensors, accelerometers, micro-cameras,
- **availability of COTS robot sensors**, including robot odometry and position, laser and cameras,
- **scenario configuration flexibility**, allowing different scenario configurations such as obstacles, nodes located at arbitrary (user-defined) locations,
- **possibility of indoor and outdoor deployments** to cover a broader range of experiments and functionalities,
- **extensibility with new hardware elements**.

    The main usability requirements were:

- **online remote operation**, the testbed should allow full remote operation including experiment execution, monitoring and logging,
- **remote experiment preparation**, the testbed should include tools to allocate and schedule testbed resources,
- **confidentiality**, the testbed should preserve intellectual and industrial properties of the methods experimented,
- **security**, secure access to the testbed and to the experiments results should be granted,
- **safety**, the testbed should include ways of ensuring safety of every device and person in the scenario,
- **easy-to-use**, it should include well documented tools, libraries with basic functionalities to facilitate the programming for non experts, code re-use, availability of operation manuals, tutorials and examples,...

- **availability of basic functions**, it should include functions to release users from the need of having to program all the devices included in their experiments allowing them concentrate on the core techniques being experimented.

This chapter describes the architecture selected for the CONET Integrated Testbed to address these requirements. The testbed usability tools, to address usability requirements, will be presented in Chap. 4.

3.3 Software Architecture

One of the main architectural requirements is the need for interaction between heterogeneous devices in an open, flexible and interoperable way. Also, it should support user programs capable of executing a wide range of different methods and experiments. Modularity, usability, extensibility and reuse of code are also to be taken into account. The solution adopted is to use an integrating layer through which all the modules intercommunicate using standardized interfaces that abstract their particularities in application programming interfaces available for a number of programming languages.

3.3.1 Integrating Layer

Figure 3.1 shows a very simplified diagram of the proposed scheme. All modules interact using an integrating software layer such that the output of one module feeds other modules in a closed-loop manner. Each of the devices to be integrated is treated as a module. This scheme can be used with single devices and with teams and networks of devices. Each module can be considered as a black box, which can be implemented indistinguishably by one or more systems, either real or simulated. This scheme supports the integration of virtual devices (either real

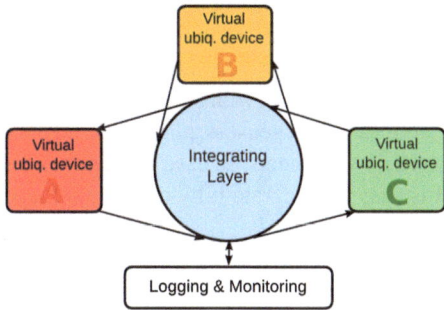

Fig. 3.1 Scheme for the integration of heterogeneous ubiquitous devices

or simulated), enabling hybrid real-simulated systems. For instance, the robot programs can be executed in real hardware, whereas the operation of the WSN is simulated. Notice that the integrating layer is an abstract software layer running in each module, not a central component. This approach also enables using different types of robots, e.g. aerial or ground robots, with diverse platforms, sensors and hardware.

Time synchronization should be carefully considered in this scheme. Different programs, devices or simulators can have very diverse computing and timing needs. Therefore, two potential approaches have been considered: synchronous and asynchronous. In the first one, each module is executed in a time slot basis in a serial way, one module after the other. This requires that the programs and simulators running in the modules have suitable and compatible timing management tools. Of course, it also implies time synchronization among modules that can be running in the same or different machines. Time slot granularity is also an issue to be taken into account.

On the other hand, in the asynchronous approach all modules are executed in parallel, without the need for a unified timing base. Each module decides depending on its needs whether to treat data requests as non-blocking operations or to block the execution of the program until the required data is received. When a module needs data from another module, the first one can be blocked until it gets the required data from the second one. In this approach there is not need for timing management or global synchronization. Nevertheless, it is necessary to take carefully into account the interactions between modules to avoid mutual blocking. This asynchronous option was adopted in the CONET Integrated Testbed.

3.3.2 System Middleware

Player[1] [4], described in Sect. 2.3, was selected as the main integrating layer of the CONET Cooperating Objects Integrated Testbed. Player has been used as middleware in a number of multi-robot testbeds due to its flexibility and modularity [5], [6]. Player is based on a client/server architecture. The Player Server interacts with the hardware elements and uses abstract interfaces to communicate with the Player Client, which provides access to all the system elements through device-independent APIs. In the CONET Integrated Testbed Player communicates on one side with the sensors, robots and sensor nodes and, on the other side, with the programs developed by users. Player includes support for a large variety of sensors, platforms and devices, making straightforward the integration of new elements. In case Player does not provide support for one specific device, it is easy to integrate it by defining a Player Driver for the new element.

---

[1] http://www.playerstage.sourceforge.net

Player is one of the predecessors of Robotics Operating System (ROS)[2]. ROS provides the services that one would expect from an operating system, and has recently gained high popularity. When the CONET Integrated Testbed was developed ROS was not still sufficiently mature while Player was widely used by an active community. Nevertheless, Player is compatible with ROS and the migration of the CONET Integrated Testbed to ROS should not require significant efforts.

3.3.3 Architecture

Figure 3.2 depicts the basic software architecture of the CONET Integrated Testbed. The figure shows the main processes that run in the robot processors, WSN nodes and in the Monitor PC. The Player Robot Servers include drivers for bidirectional communication with the low-level robot controller and the most frequently used sensors such as cameras, laser range finders and also with a WSN node that is mounted on each robot. The Player WSN Server runs in the Monitor PC, which is connected to a WSN gateway. The WSN driver also runs in each of the robots to communicate with their on-board WSN node.

This architecture allows several degrees of centralization. In a decentralized experiment, programs are executed on each robot and, by means of Player Interfaces, they have access to the robot local sensors. Also, each Player Client can access any Player Server through a TCP/IP interprocess connection. Thus, since robots are networked, the Player Client of one robot can access the Player Server of each robot, as shown Fig. 3.2. In a centralized experiment a Central User Program can connect to all the Player Servers and have access to all the data of the experiment. Of course, centralized schemes have bad scalability and in some cases they can be significantly constraint. In any case, these centralized approaches can be of interest for debugging and development purposes. In the figure the Central User Program is running in the Monitor PC. It is just an example. It could be running, for instance, in one robot processor. Following this approach, any other program required for an experiment can be included in the architecture. To have access to the hardware, it only has to connect to the corresponding Player Server and use the Player Interfaces.

A Graphical User Interface (GUI) was developed to provide remote users with full on-line control of the experiment including programming, debugging, monitoring, visualization and logs management. It works as yet another Player Client and connects to all available Player Servers, gathering data of interest of the experiment. This GUI will be presented in Chap. 4 together with other usability tools that significantly reduce the experiment development and debugging efforts. Note that the architecture assumes that all its modules are either physically or wireless connected allowing remote operation over the Internet.

---

[2] http://www.ros.org

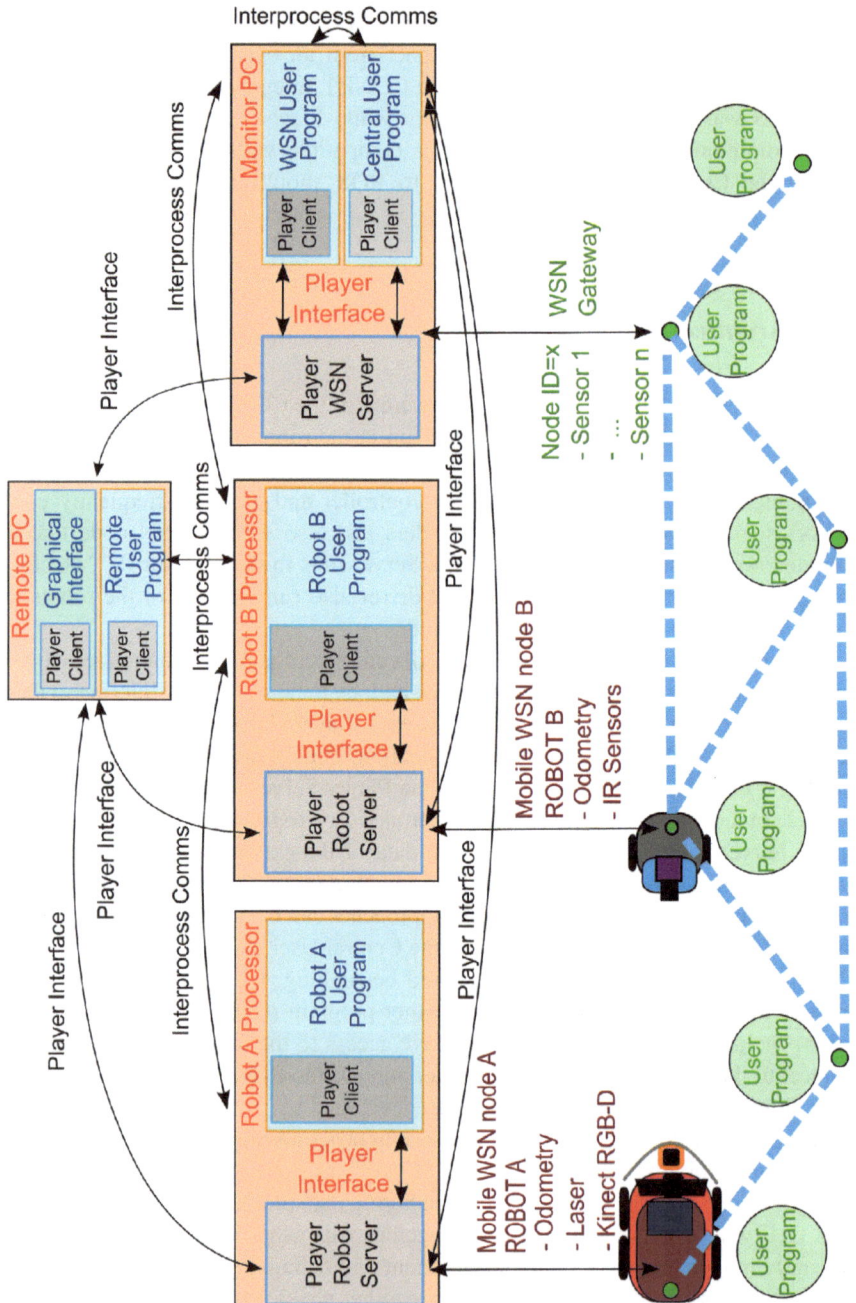

Fig. 3.2 General scheme of the architecture of the CONET Cooperating Objects Integrated Testbed

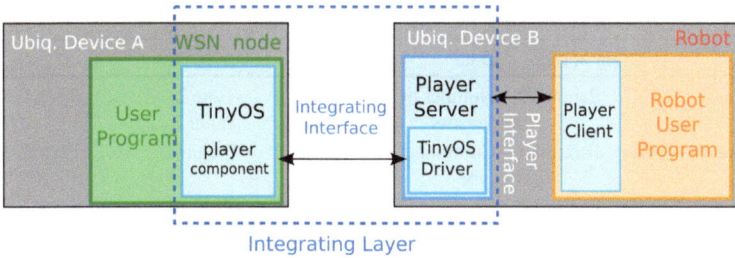

Fig. 3.3 Scheme for peer-to-peer interoperability between WSN nodes and robots in the CONET Integrated Testbed

Figure 3.2 shows in blue color the modules provided as part of the CONET Integrated Testbed infrastructure. The user programs to be experimented should be implemented in these modules: WSN programs in green color; robots, central and remote programs in orange. The system also includes tools to facilitate experimentation, such as a set of commonly-used basic functionalities for robots and for WSN that prevent the user to necessarily have to program all devices. These functions allow him to concentrate on the experiment itself. They will be described also in Chap. 4.

3.3.4 Interoperability

To allow peer-to-peer interoperability between WSN and robots an interface between WSN nodes and Player has been developed. Its objective is to integrate WSN within Player so that the internal behavior of the WSN including messages formats, protocols, programming language and WSN operating system are transparent to the rest of the system, see Fig. 3.3. It gives users freedom to design their WSN and multi-robot internal behaviors and protocols with the only constraint that the messages between the WSN and the robots should comply with the defined interface. The objective is to specify a common language between ubiquitous devices and, at the same time, give flexibility to enable a high number of experiments.

The developed interface allows communication among individual devices (e.g. robot with its connected WSN node) and among teams or networks (e.g. Monitor PC with the entire WSN gateway or team of robots). The same interface is used for communication between individual WSN nodes (or the WSN as a whole using a gateway) and individual robots as well as for communication between individual WSN nodes (or the WSN as a whole using a gateway) and the team of robots as a whole.

This interface is comprised of bidirectional messages consisting of a header with routing information and a body, which depends on its type. General data messages, requests and commands are considered for both, incoming and outgoing messages. Some usual application-dependant message types such as Radio Signal Strength

Table 3.1 Examples of messages in the WSN-Player interface

Type	Routing header		Data				
Sensor data	CO ID	Parent ID	Number of sensors	Type 1 value 1	Type 2 value 2	...	Type N value N
Command	CO ID	Parent ID	Command type	Param. size	Param. 1	...	Param. N
Position	CO ID	Parent ID	X	Y	Z		State
User data	CO ID	Parent ID	Data size	Byte 1	Byte 2	...	Byte N

Indicator (RSSI), sensor data and alarms were defined as well. Any WSN comply-
ing with this interface can be straightforward integrated in the CONET Integrated
Testbed. Table 3.1 shows the format of these messages.

Using this interface robots can request measurements from the WSN node they
carry. In an active perception experiment, the robot can command the WSN node
to deactivate a sensor if the measurements it provides have no suitable information
content. Also, a WSN node can command the robot to move to a certain location in
order to improve its perception. Each robot can communicate not only with the WSN
node it carries, but also with any other node in the WSN. Thus, any robot can request
the readings from any WSN and any WSN node can command any robot. For instance,
in a robot-WSN data mule experiment one node could command a robot to approach
a previously calculated location. Also, this robot-WSN communication can be used
for sharing resources: a WSN node can send data to the robot in order to perform
complex computations or to register logs benefiting from its higher computational
capacities.

The implementation of this WSN-Player interface required the development of
a new Player module. Also, a new TinyOS component was developed providing a
transparent API compliant with this protocol in order to facilitate program development. The component was validated with Crossbow TelosB, Iris, MicaZ, Mica2
WSN nodes.

This Player module has been approved by Player developers for integration in the
next version of Player 3.1.x and can be downloaded, integrated in Player, from [7].

3.4 Hardware

Figure 3.4 shows the basic deployment of the CONET Cooperating Objects Integrated
Testbed. It is set in a room of more than $500\,m^2$ ($22\,m \times 24\,m$) with three columns.
Two doors lead to a symmetrical room to be used in case extra space is needed at
the basement of the main building of the Shool of Engineering of the University of
Seville. The figure shows the mobile robots and WSN nodes (green dots), that can be
mobile when mounted on the robots or carried by people. It also includes a network
of static cameras (in yellow).

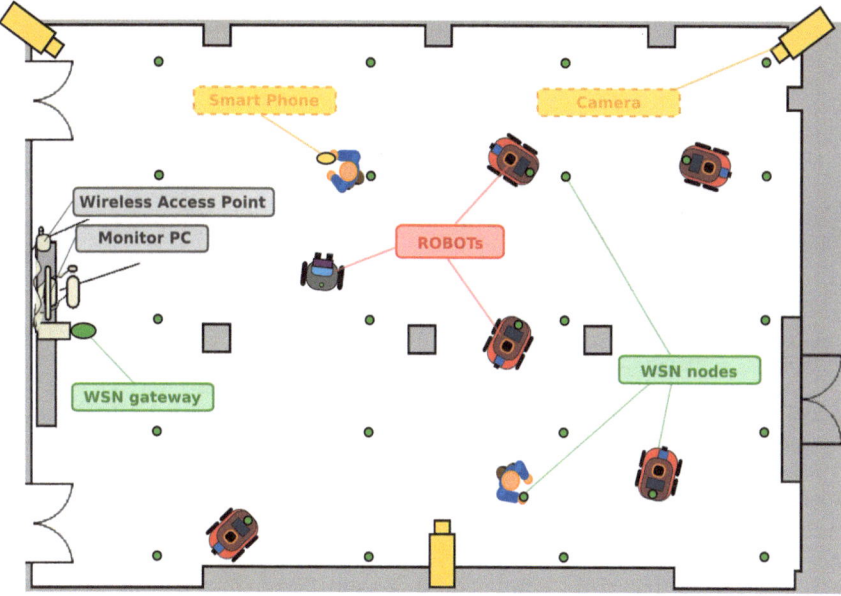

Fig. 3.4 General scheme of the CONET Integrated Testbed

Three main hardware components were taken into account: platforms (static and mobile), sensors and communication infrastructure. Two main platforms are used: mobile robots and Wireless Sensor Networks, with high differences in their sensing, computing and communication capabilities.

3.4.1 Mobile Robots

The testbed currently includes 5 skid-steer holonomic Pioneer 3-AT robots manufactured by Mobilerobots.[3] The basic Pioneer 3-AT platform was enhanced with several sensors, extra computational resources—a Netbook PC with an Intel Atom processor and 1024 MB SDRAM—and communication equipment—a IEEE 802.11 a/b/g/n Wireless bridge, see Fig. 3.5-left. The main sensors in each robot are a Microsoft Kinect, a Hokuyo laser range finder and one WSN node serially connected to the robot, which enables robot-WSN cooperation and provides extra sensing capabilities. Each robot can also be equipped with a RGB IEEE1349 camera, in case robots with two cameras are required in an experiment. Each robot is equipped with one GPS card and one Inertial Measurement Unit (IMU) for use in outdoor experiments.

The testbed also includes one Rain robot developed at the *Dept. Ingegneria della Informazione* of University of Pisa, Fig. 3.5-right. This is a low-weight robot with differential configuration. It includes a TMoteSky node that is used as the main

---

[3] http://www.mobilerobots.com

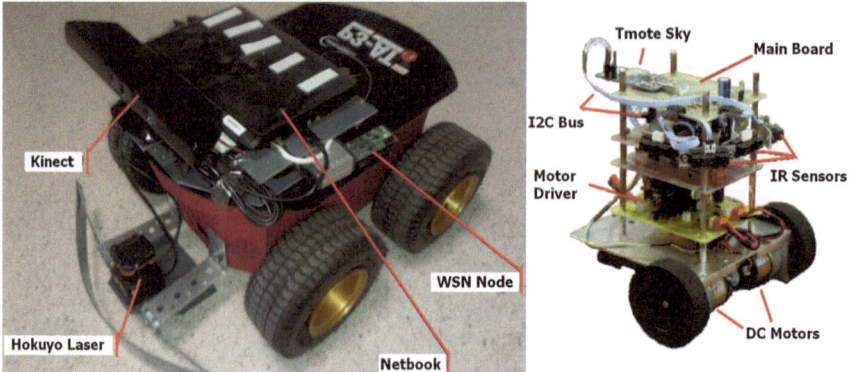

Fig. 3.5 Robots currently used in the CONET Integrated Testbed: five Pioneer 3-AT (*left*) and one Rain robot (*right*)

processor of the robot. The robot includes a micro-controller responsible for low-level motion control. Messages to the robot are encapsulated in frames. The robot node interprets the messages and sends the corresponding commands to the micro-controller through I2C. The robot is equipped with wheel odometry sensors and five infrared distance sensors for collision detection.

3.4.2 Sensor Networks

The sensor network of the Cooperating Objects testbed consists of static and mobile nodes. Each robot is equipped with a WSN node. Nodes can also be carried by people if required in an experiment. The static WSN nodes are deployed at 21 predefined locations hanging at 1.65 m height from the floor. A diagram of the network deployment is shown in Fig. 3.6. This configuration allows establishing WSN node clusters in order to enable WSN clustering experiments by switching off some of the nodes (e.g. node n10).

A dedicated processor (Monitor PC) is physically connected to all WSN nodes through a USB-hub tree. The Monitor PC allows monitoring, programming and logging the WSN nodes. It also connects the WSN with the rest of the system elements. There are two ways of communicating WSN nodes and the Monitor PC: wired and wireless connections. Using both can be interesting for instance to separate different types of traffic. The testbed allows freedom to use any, both or none of them in the experiments.

Six different models of WSN nodes are currently available at the CONET Integrated Testbed, four from Crossbow[4]: TelosB, Iris, MicaZ and Mica2, and two from Advanticsys[5]: CM5000 and CM5000-SMA, see Fig. 3.7. TelosB, CM5000 and

---

[4] http://www.xbow.com

[5] http://www.advanticsys.com

CM5000-SMA are equipped with SMD (Surface Mounted Devices) sensors whereas the rest need to be equipped with MTS400 or MTS300 sensor boards. Also, some have been equipped with embedded cameras such as CMUcam2 and CMUcam3.[6] A detailed description of the sensors currently available is provided in Sect. 3.4.4.

3.4.3 Communications

Two wireless networks are available at the CONET Integrated Testbed: a Wi-Fi LAN (IEEE 802.11b/g/a) that links the Monitor PC and the team of robots and an ad-hoc network used by the WSN nodes. Robot and WSN networks differ greatly in range, bandwidth, quality of service and energy consumption. The IEEE 802.15.4 standard has a bit rate of 250 kbps and a range of less than 40 m in realistic conditions. On the other hand, Wi-Fi can provide up to 54/36 mbps (maximum theoretical/experimental bound) at significantly larger ranges.

Two different WSN networks can be used in the CONET Integrated Testbed. TelosB, Iris, MicaZ, CM5000 and CM5000-SMA use a network based on IEEE 802.15.4, whereas Mica2 nodes use an ad-hoc protocol that operates in the 900 MHz radio band. Both WSN networks can be used in the same experiment. We installed a dedicated Wi-Fi network operating at 5 GHz in order to avoid potential interference with IEEE 802.15.4 and with the 2.4 GHz Wi-Fi network of the School of Engineering.

3.4.4 Sensors

A rich variety of sensors have been integrated in the CONET Integrated Testbed. Table 3.2 shows the main features of the sensors mounted on the mobile robots. Besides the sensors in the table, one magnetic compass was integrated in each robot for use mainly in outdoor experiments. Compasses provide measurements at 10 Hz with a resolution of 0.08°. To enlarge the range of the experiments all sensors mounted on the robots are suitable for outdoor experiments, with the exception of the Kinect distance sensor, which is affected by solar light.

Table 3.3 shows the features of the main WSN sensors used in the CONET Integrated Testbed. The WSN nodes also include sensors to measure the strength of the radio signal (RSSI) and Link Quality Indicator (LQI). Each model measures RSSI differently due to different antennas and radio circuitries. For instance, MicaZ uses an 1/4 wave dipole antenna with −94 dBm sensitivity, TelosB nodes use Inverted-F μstrip antenna with −94 dBm sensitivity, and Iris nodes also uses an 1/4 wave dipole antenna with −101 dBm sensitivity. Other WSN nodes integrate Figaro gas concentration detectors, infrared barrier sensors and smoke detection sensors.

---

[6] http://www.cmucam.org

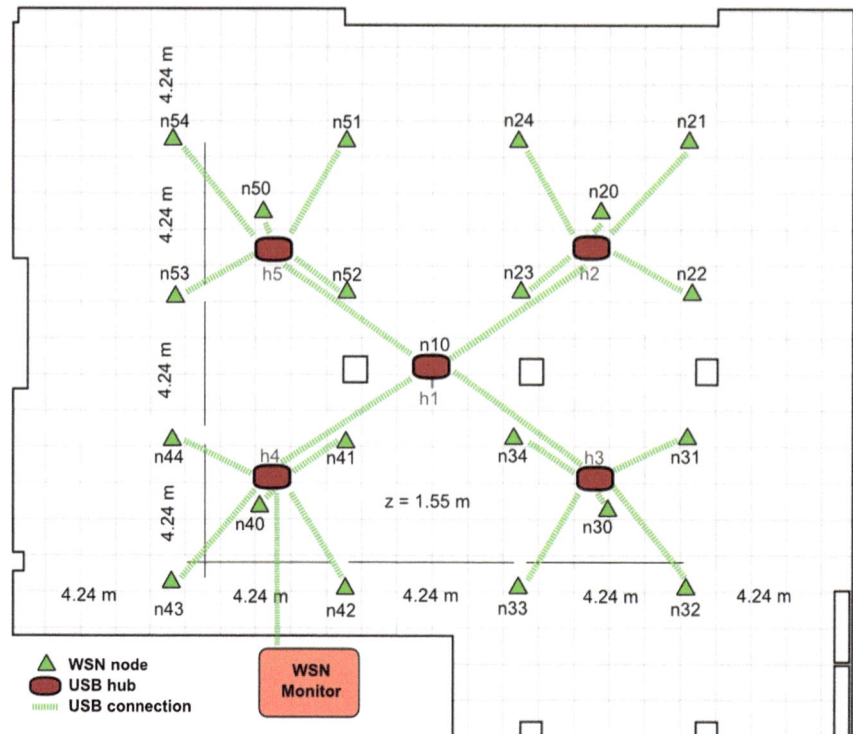

Fig. 3.6 Deployment of the static WSN in the testbed room

Fig. 3.7 WSN nodes deloyed in the CONET Integrated Testbed: (*left*-to-*right*) Crowbow Mica2, Telosb, Micaz, Iris and Advanticsys CM5000 and CM5000-SMA

Table 3.2 Main features of the sensors mounted on the mobile robots used in the CONET Integrated Testbed

Sensor	Physical magnitude	Main specifications	Power (W)
Hokuyo UTM-30LX	Distance (m)	Range 0.1–30 (m) Accuracy ±3 (cm) FOV 270 (°) Resolution 0.25 (°)	8
Microsoft kinect	Distance (m) Color (RGB) Infrared image Accelerat. (m/s²)	Range 0.4-4 (m) Resolution 640 × 480 (px) FOV = (57,43) (°) Freq. 30 (fps)	12
Imaging source 21BF04 camera	Color (RGB) Light (intensity)	Resolution 640 × 480(px) Freq. 60 (fps)	2.4
MC-1513 GPS	Longitude (°) Latitude (°) Altitude (m) Velocity (m/s)	Accuracy 2.5 (m) Freq. 10 (Hz) Range 0–18,000 (m) Velocity range 0–515 (m/s)	0.2
Ez-compass-3A	Angle (°) Magnetic field (Gauss) Accel. (m/s²)	Accuracy 0.5 (°) Resolution 0.08 (°)	0.54

Table 3.3 Main features of WSN sensors used in the CONET Integrated Testbed

Sensor	Physical magnitude	Main specifications	Power (mW)
CMUcam2/3	Color image	Resolution 352 × 288 (px)	650
		Freq. 1 (fps)	
Hamamatsu S1087	Light intensity	Spectral range 320–730 (nm)	<1
	Infrared intensity	Spectral range 320–1100 (nm)	
Sensirion SHT11	Humidity (%)	Humidity range: 0–100 (%)	<1
	Temperature (°C)	Temperature range: −40 to 123.8 (°C)	
		Accuracy ±3.5 (%) (25°C)	
Panasonic ERTJ1VR103J	Temperature (°C)	Range −40 to 70 (°C)	<1
CdSe photocell	Light intensity	Maximum sensitivity 690 (nm)	<1
		Range 2–520 (kΩ)	
Panasonic WM62A	Sound frequency (Hz)	Range 20–16000 (Hz)	<1
Microphone	Amplitude (V)	Sensitivity −45 ± 4 dB	
Intersema MS5534AM	Pressure (mbar)	Range 300–1100	<1
		Resolution 0.01	
		Accuracy ±1.5 % (25°C)	
ADXL202JE 2-axis accelerometer	Acceleration (m/s²)	Range ±2(g) Resolution 2(mG)	<1
		Sensitivity 167 (mV/G)±17 %	
		Accuracy 2(mg) (60 Hz)	
Honeywell HMC1002	Magnetic field (Gauss)	Resolution 27 (μgauss)	20

Combining robot and WSN sensors with diverse physical, computation and communications characteristics can be very interesting in Cooperating Objects experiments. WSN sensors usually have lower sensing capabilities (accuracy, sensitivity, resolution), lower output bandwidth in order to simplify the transmission and processing of the measurements and lower energy consumption than those carried by robots.

The RGB cameras in the robots and those used by the WSN (CMUcam3) are a clear example of this difference. The former have a resolution of 640x480 pixels, a frame rate of 30 frames per second and a power consumption of 2.4 W. On the other hand, the latter provides images of 352×288 pixels at 115200 bits per second (around 1 frame per second) but only consumes between 135 and 650 mW.

3.5 Conclusions

This chapter presents the architecture of the CONET Cooperating Objects Integrated Testbed. The software architecture and hardware components were developed following the guidelines and requirements inferred from the questionnaires filled in by potential users from industry and academy.

Interoperability, flexibility and usability have been the main requirements for its software architecture. The solution adopted is to use an integrating layer through which all the modules intercommunicate in a peer-to-peer basis using standardized interfaces. The hardware components include a WSN with static and mobile nodes, a camera network and a fleet of mobile robots, communication infrastructure and sensors that are frequently used in multi-robot, WSN and Cooperating Objects research and experimentation.

Special interest was devoted to the usability of the CONET Integrated Testbed. A number of tools and functions were developed to allow secure, remote, easy-to-use testbed operation. These tools are described in the next chapter.

References

1. Ramiro Martinez-de Dios J et al., (2011) El Libro Blanco de la Robótica en España. White Book of Robotics in Spain, CEA
2. Banâtre M, Marrón PJ, Ollero A and Wolisz A, editor (2008) Cooperating Embedded Systems and Wireless Sensor Networks. ISTE Wiley, London, UK. ISBN 978-1-84821-000-4
3. Marrón PJ, Karnouskos S, Minder D and the CONET Consortium, editor (2009) Research Roadmap on Cooperating Objects. Office for Official Publications of the European Communities, Luxembourg. ISBN 978-92-79-12046-6
4. Gerkey BP, Vaughan RT, Howard A (2003) The player/stage project: tools for multi-robot and distributed sensor systems. In: Proceedings of the international conference on advanced robotics, pp 317–323
5. Michael N, Fink J, Kumar V (2008) Experimental testbed for large multirobot teams. IEEE Robot Autom Mag 15(1):53–61
6. Riggs T, Inanc T, Zhang W (2010) An autonomous mobile robotics testbed: construction, validation, and experiments. IEEE Trans Control Syst Technol 18(3):757–766
7. Player. Current Player Distribution. http://sourceforge.net/projects/playerstage/files/

Chapter 4
Usability Tools

4.1 Introduction

Usability is an important requirement in a testbed developed to bridge between different research communities. The CONET Cooperating Objects Integrated Testbed should be easy to use to potential users interested in WSN, networked robots and robot-WSN cooperation fields. The testbed should allow full remote control over the experiment and its operation should be simple, reliable, robust and secure. This chapter presents the main usability tools that have been made available to the testbed users.

This chapter is structured as follows. Section 4.2 describes the main infrastructure to enable remote operation, including the Graphical User Interface to remotely program and control the experiment and the CONET Integrated Testbed website for experiment scheduling and resources allocation. Section 4.3 describes the set of basic functionalities currently included in the testbed infrastructure and available to users. Section 4.4 describes the characterization of the testbed room. Section 4.5 illustrates the integrated simulation possibilities. Finally, Sect. 4.6 concludes the chapter.

4.2 Accessibility

The possibility of performing remote experiments is critical when considering a testbed as a tool to serve a community. In the CONET Integrated Testbed remote access is achieved by means of a Virtual Private Network, specifically set up with authentication and ciphering to ensure secure communications over the Internet. Also, remote experiment execution requires tools for experiment scheduling, execution and results registering and downloading. On one hand, this requires the development of a web site where registered users can schedule experiments and allocate resources before the experiment. On the other, a Graphical User Interface (GUI) has been developed

J. R. Martinez-de Dios et al., *A Remote Integrated Testbed for Cooperating Objects*,   41
SpringerBriefs in Cooperating Objects,
DOI: 10.1007/978-3-319-01372-5_4, © The Author(s) 2014

for interaction with the testbed during the experiment for programming, control of the experiment, visualization of the experiment data, results logging and downloading.

4.2.1 Virtual Private Network

As is known, a Virtual Private Network (VPN) is a computer network in which some of the links are carried out by open connections or virtual circuits in some larger networks (such as the Internet). One common application is to secure communications through a public network. The second main benefit of VPNs is the ability to separate the traffic of different users over an underlying network with specific security features, or to provide access to a network via customized or private routing mechanisms. VPN allows remote users to connect to the testbed securely and in a seamless way, simplifying system setup and configuration.

The testbed VPN is used to provide secure network connections between the CONET Integrated Testbed and its remote users. Each registered user has its own keys, which were sent to him after registering. OpenVPN[1] was the VPN software selected for the CONET Integrated Testbed. An authenticated user connects to the VPN Server using a secure SSL channel. Once the connection is established, the VPN server will automatically provide the network configuration. This information is sent to the VPN client in the user's computer and is used to properly configure the virtual connection. All these steps are transparent to the user. Once the VPN connection has been established, the user has access to the CONET Integrated Testbed components and tools as if being physically at the testbed LAN.

The VPN Server is located at the Laboratory Buildings of the School of Engineering and has a dedicated optical fiber cable that connects it to the testbed infrastructure located at the basement of the main building of the School of Engineering. This ensures quality Internet access. In the experiments carried out we noticed that the testbed user should have a nominal bandwidth of around 10 MB/s to enable receiving simultaneously camera images and laser readings with no significant delays. This constraint has low practical impact and testbed users can connect to the CONET Integrated Testbed from Wi-Fi public networks with moderate traffic levels. Moreover, these bandwidth requirements could be modified adapting the implementation of the experiment. If the experiment is implemented locally, avoiding executing of bandwidth demanding methods in the Remote PC (recall Fig. 3.2), it would be necessary only to transmit over the Internet data and images for experiment visualization and monitoring. Their rate and resolution could be made lower by configuring Player Servers accordingly.

---

[1] http://openvpn.net

4.2.2 Website

The CONET Integrated Testbed website (http://conet.us.es) is the primary interface between the users and the manager of the CONET Integrated Testbed, the testbed engineer. The website provides all necessary documentation, software, tutorials and examples. It also allows experiment scheduling and resource allocation as well as post-experiment results and logs download, see Fig. 4.1. It also keeps a log with all experiments requested, scheduled and performed by the user.

Figure 4.2 shows the flow of interactions between users and the testbed engineer. To request an experiment, after registering the user should allocate the resources (such as number of robots, number and types of WSN sensors) needed for the experiment and the desired date and hour for the experiment. The resource allocation request also includes a brief description of the experiment and any specific robot or scenario configuration, such as obstacles for robot motion. The testbed engineer will confirm the feasibility of the experiment and the availability of resources. After confirmation, the experiment will be performed as agreed.

During the experiment the user will have full control of the execution by means of the CONET Integrated Testbed GUI. The testbed engineer will attend the experiment locally for safety reasons and to provide support if needed. Prior to the experiment, users should test the method with the simulation tools described in Sect. 4.5.

Access to the private part of the website requires being registered. This is convenient in order to maintain an updated list of users and to restrict the access to the testbed if needed. After registration the user obtains website restricted area access and the corresponding VPN connection keys.

A Virtual Machine with all necessary software and pre-installed libraries is available to facilitate the software development and installation process. The Virtual Machine currently available at (http://conet.us.es) includes: Ubuntu OS, Player, the

Fig. 4.1 Snapshot of the CONET Integrated Testbed website (http://conet.us.es)

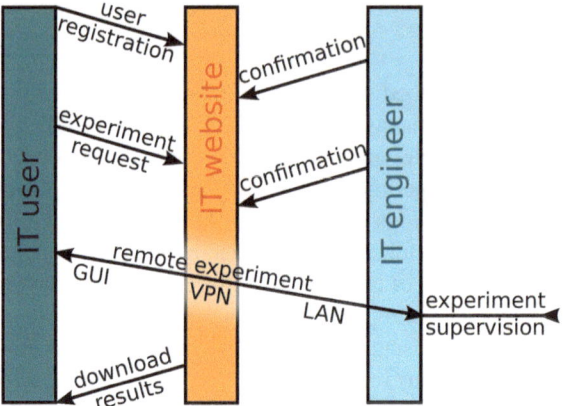

Fig. 4.2 Flow of interactions for experiment scheduling between users and the testbed engineer

CONET IT Graphical User Interface, TinyOS 1.x and 2.x, the WSN-Player interoperability TinyOS component as well as documentation and tutorials.

4.2.3 Graphical User Interface

The CONET Integrated Testbed GUI, in Fig. 4.3, has been developed to facilitate the remote use of the testbed. Its main functions include programming, monitoring and controlling the experiment. The GUI is fully integrated in the testbed software architecture and allows remote access to all the devices using Player Interfaces. The GUI can be used to monitor the experiment, including the position and orientation of the robots and readings from the WSN sensors. It contains tools to visualize images and laser readings from the robots. Images from the camera network can be also used to remotely visualize the experiment.

The CONET Integrated Testbed GUI also allows remotely programming and reprogramming each of the elements involved in the experiment, robots or WSN nodes. This includes configuring and running basic functionalities for each type of platform that will be presented in Sect. 4.3. For instance, the robot trajectory following functionality can be configured by simply providing a list of waypoints. The waypoints can be given by manually writing the coordinates in the dialog box or by means of a text file. Also, the user can graphically (by clicking on the GUI window) define the robot waypoints of the trajectory. The GUI can be used to upload user executable codes for each element involved in an experiment. The user uploads executable codes to preserve confidentiality and intellectual property of the method. It is also possible to re-program robots and/or WSN nodes in between experiments facilitating debugging.

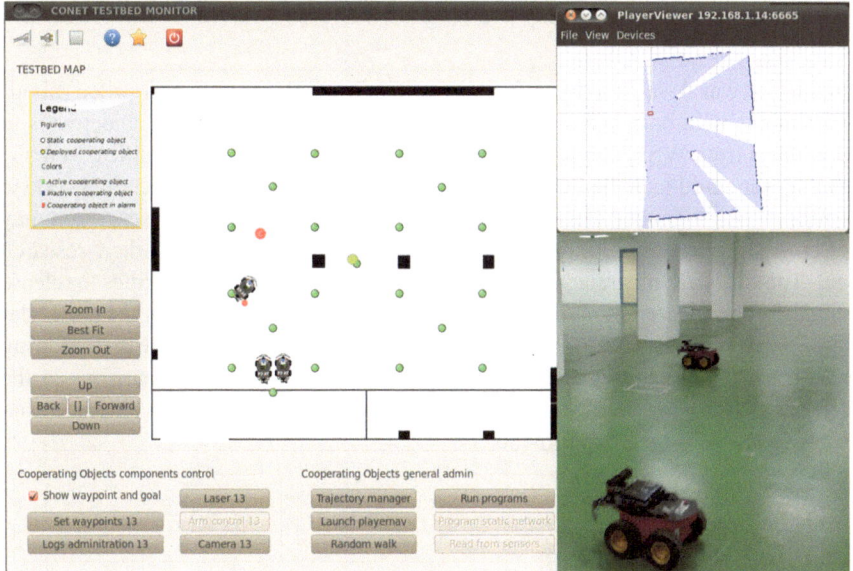

Fig. 4.3 Snapshot of the CONET Integrated Testbed GUI during an experiment

The CONET Integrated Testbed GUI also allows full control of the experiment. It is possible to start/stop the programs involved in one experiment synchronously or individually one by one. Notice that WSN nodes programs will start running just after programming its EPROM. Thus, it is recommended to synchronize WSN nodes by an initial packet sent for instance by an onboard WSN node commanded by a robot. Finally, the GUI offers remote logging control, allowing the user to start or stop logging. To cope with potential bandwidth limitations of such a remote access, the user can select in the GUI the data he wants to monitor and log, their rate and resolution.

Below are the main functions implemented in the CONET Integrated Testbed GUI:

- monitoring of WSN packets, data and alarms,
- monitoring of robot locations, orientations and data,
- visualization of live images and laser readings from each robot,
- programming of WSN and robots individually or all together,
- robot basic functionalities: obstacle avoidance, random walk, waypoint following using a list of waypoints, a text file or clicking on screen,
- WSN basic functionalities: data collection using the Collection Tree Protocol (CTP) [1],
- general experiment start/stop,
- start/stop of individual robot programs,
- logging start/stop of each individual robot,
- logging start/stop of WSN.

4.3 Basic Functionalities

The testbed was developed to perform experiments involving only networked robots, experiments with only WSN and experiments integrating both. The objective is to serve users from WSN, multi-robot and Cooperating Objects communities. In many cases a user could lack the background to be able to provide fully functional code to control all heterogenous elements involved in an experiment. Also, users often may not be willing to concentrate on details of techniques from outside their research field. The CONET Integrated Testbed includes a set of basic functionalities to release the user from programming the modules that may be unimportant in his particular experiment, allowing him to concentrate on the algorithms being tested. Sharing these basic functionalities among users enables more accurate comparisons of the methods performed. Below are some basic functionalities currently available and usable by users in their experiments.

4.3.1 Indoors Positioning

Outdoors localization and orientation of robots is carried out with GPS and Inertial Measurement Units. For indoors settings a beacon-based computer vision system is used. Cameras installed on the room ceiling were discarded due to the number of cameras—and computer power to carry out image processing methods—required to cover our $500\,\mathrm{m}^2$ environment. In the solution adopted each robot is equipped with a calibrated webcam pointing at the room ceiling, on which visual beacons have been stuck at known locations. The beacons are distributed in a uniform square grid: the webcam always sees in the image at least one beacon square, i.e. an imaginary square with one beacon at each corner.

Figure 4.4 summarizes the main steps in the localization vision-based method implemented in the CONET Integrated Testbed. Each robot processes the images from its webcam: segments the beacons in the image, computes the beacon centroid on the image plane, extracts its geometry—area and eccentricity—and classifies the beacon using a minimum distance classifier. Beacons were selected to allow their classification using simple and rotation-invariant features.

Assume that the cameras have been previously internally calibrated. Then, analyzing the beacon types that form the square, the beacon square is univocally identified. The beacons were deployed such that each beacon square is unique. Thus, it is possible to recognize the beacon square and therefore each of its individual beacons. The coordinates of each beacon on the ceiling is known. Thus, using the coordinates of the beacons identified it is possible to efficiently compute the robot location and orientation applying efficient homography-based techniques. Assuming that the robot reference system has null roll and pitch angles, the robot location and orientation can be computed using two beacons in the image. If more than two beacons are seen in the image, higher accuracy can be obtained using methods based on least squares.

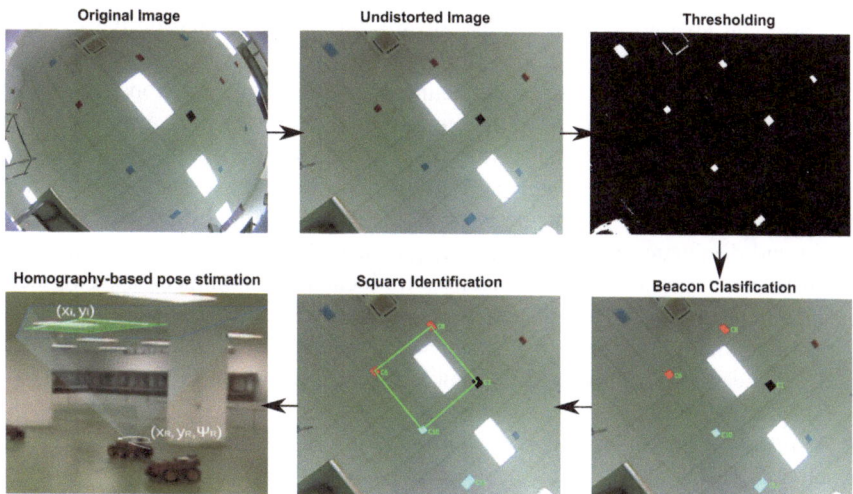

Fig. 4.4 Main steps in the vision-based localization method used in the CONET Integrated Testbed

The implemented localization method has been validated experimentally with a mean error lower than 7 mm and standard deviation lower than 5 mm, sufficient for the envisioned experiments. Moreover, the method is computationally efficient and can be executed at suitable frame rates—10 frames per second—in the robots processors, where other processes are running simultaneously.

The testbed offers the possibility to use other localization methods, such as the Adaptive Monte-Carlo Localization method (AMCL) [2]. This method maintains a probability distribution of the robot poses using a Particle Filter (PF) that adjusts its number of particles, balancing between processing burden and localization accuracy. The method uses robot odometry in the PF prediction stage. The PF update stage corrects predictions matching the map of the scenario—assumed known—with measurements from the laser range-finder.

The operation of the localization method is transparent to the user. If the method, based on vision or AMCL, is active the estimations of the robot location and orientation are used by the GUI for robot pose visualization and logging.

4.3.2 Networked Robots Basic Functionalities

Several robot motion control functionalities are offered as basic functionalities including: low-level velocity control, local position control, trajectory following and random walk.

Robot navigation basic functionalities integrated in the CONET Integrated Testbed allow the robot to follow a trajectory specified with a set of ordered waypoints.

The user can give waypoints in a list, or in a file or by clicking on the CONET Integrated Testbed GUI.

The testbed includes a global path planner based on Wavefront propagation. The method assumes that a cell-discretized map of the environment is available. Each cell is assigned with a cost that is inversely proportional to the distance to the nearest obstacle. When the planner is given a destination, it searches a path by assigning costs to adjacent cells. The method selects the cells with the lowest cost between origin and destination and identifies straight lines between these cells. The final points in these lines are used as references for local navigation methods. Two local navigation methods with automatic obstacle avoidance are implemented: the Vector Field Histogram plus VFH+ [3], suitable for robots with differential motion such as the Pioneer 3-AT robots, and the Smooth Nearest Diagram (SND) method [4], suitable for Ackerman-configured vehicles.

A module for patrolling the testbed scenario using several robots has also been developed. This module divides the testbed scenario in areas to be surveyed by each of the robot. The division takes into account the number of robots available, the robot speed and the sensor coverage in order to minimize the patrolling time. Then, the method computes waypoints for each robot and generates waypoint lists that are given to the global path planner of each robot.

A random walk basic functionality was developed by commanding the robot with pseudo-random velocity signals. Also, the underlying obstacle avoidance function from Player is adopted to ensure a minimum configurable distance with obstacles detected with the laser range finder. For safety reasons, obstacle avoidance is by default configured active in each experiment but it can be deactivated on demand if needed.

4.3.3 WSN Basic Functionalities

The Cooperating Objects Integrated Testbed also includes modules to collect, register and log in the Monitor PC the readings gathered by the sensors of static and mobile WSN nodes. This functionality configures the static WSN nodes to periodically read from their sensors and send the data to the WSN base using the Collection Tree Protocol (CTP) algorithm [1].

CTP is an efficient multi-hop routing protocol for transferring data from one or more static nodes to one or more root nodes. CTP relies in Link Quality Estimators to dynamically select the route and can quickly adapt to the changes in the network topology. CTP is adaptable and efficient in the sense that it includes mechanisms that do not require too much communication among the nodes.

CTP operates only with static nodes. Two different alternatives have been implemented to collect data from WSN nodes on robots: use the robot network or use the routing channels of the WSN static network. In the first case, the messages are sent to the corresponding robot using the WSN-Player interface. The robot then forwards the data to the Monitor PC using the multi-robot Wi-Fi network. In the second case, the mobile WSN node sends the data to a static node that forwards the packet to the gateway using static WSN routing channels. The mobile node periodically broadcasts beacons asking for responses from static WSN nodes in order to select the static node within its radio coverage with which he has the best link quality.

The Cooperating Objects Integrated Testbed also includes static WSN functionalities for network formation using Xmesh. Xmesh is a distributed routing method based on the minimization of a cost function that considers link quality of nodes within a communication range [5].

4.3.4 Synchronization

Sensor data synchronization is required in a wide range of experiments. The solution adopted is to use time stamps. The robots and the Monitor PC, connected through Wi-Fi, are synchronized using the well-known Network Time Protocol (NTP) [6]. For the WSN nodes we implemented the Flooding Time Synchronization Protocol (FTSP) [7]. The FTSP leader periodically sends a synchronization packet with its local time. Each node that receives the message re-sends it following a flooding strategy. The local time of each node is corrected using that of the FTSP leader depending on the time stamp of the packet and on the sender of the message. The algorithm is efficient and achieves synchronization errors of few milliseconds, sufficient for a wide number of applications.

In our case the FTSP leader node is the WSN base. The Monitor PC, connected to the WSN base, is synchronized with the robots using NTP. The Monitor PC uses NTP time as reference to the FTSP leader. Thus, all robots and WSN nodes are synchronized.

4.4 Testbed Room Characterization

A set of experiments were carried out to characterize the testbed room in order to facilitate experimentation.

4.4.1 Radio Channels Characterization

The testbed is equipped with two WSN sniffers for network surveying. The first monitors the use of every channel in the 2.4 GHz band. The second registers all packets interchanged in the WSN network. We used these sniffers to measure the level of interference of the different IEEE 802.15.4 channels, see Fig. 4.5. The CONET Integrated Testbed provides the possibility to select the degree of intensity on the WSN channel interference. For instance, in preliminary experiments it can be interesting to

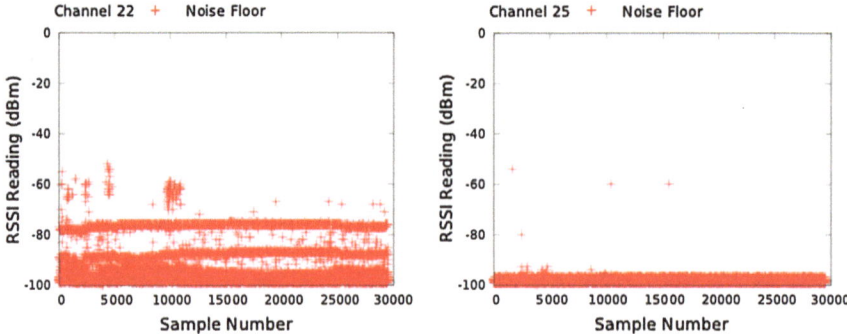

Fig. 4.5 Interference level in different IEEE 802.15.4 channels: *left* channel 22 with high interference; *right* channel 25 with low interference level

select channels with low interference such as channel 25, see Fig. 4.5. On the other hand, when performing robustness analyses, it can be interesting to use channels with intense interference, such as channel 22.

4.4.2 RSSI-Range Model

GPS-denied localization and tracking has attracted high interest in WSN research. Many of these methods are based on RSSI. A set of experiments were carried out to obtain a RSSI-range model of the CONET Integrated Testbed. Figure 4.6 shows the RSSI-range model obtained using RSSI measurements between each pair of static nodes in the testbed room. Figure 4.7 shows the RSSI readings taken from three different nodes to the same receiver during a total of 16 h (20,000 measurements). These experiment were repeated in different days with different conditions including during night and weekend in order to assess the variability depending on the level of traffic in the 2.4 GHz band. This result revealed that RSSI measurements in the testbed were rather stable along time which favors consistent radio channel characterization.

4.4.3 RSSI Maps

A RSSI map of the testbed room has also been obtained. They are of interest for RSSI-based localization techniques such as fingerprinting. These maps are provided to the testbed users releasing them from the need of generating them.

The maps were generated by robots patrolling the scenario. Periodically the WSN node onboard each robot transmitted a beacon message. Static WSN nodes receiving the message measured its RSSI and transmitted a respond message. Thus, for each location, the robot nodes received two measurements from each static node it had

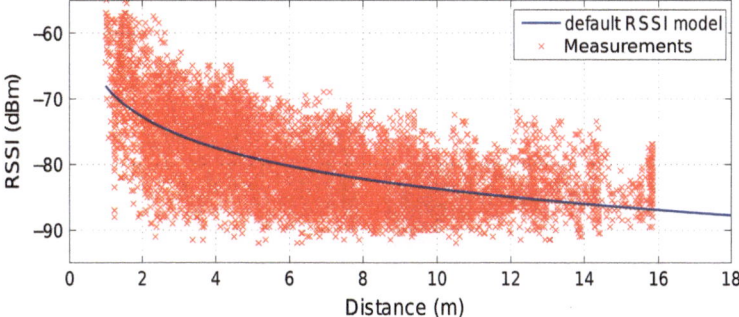

Fig. 4.6 RSSI-range model obtained using measurements between each pair of static WSN nodes in the CONET Integrated Testbed

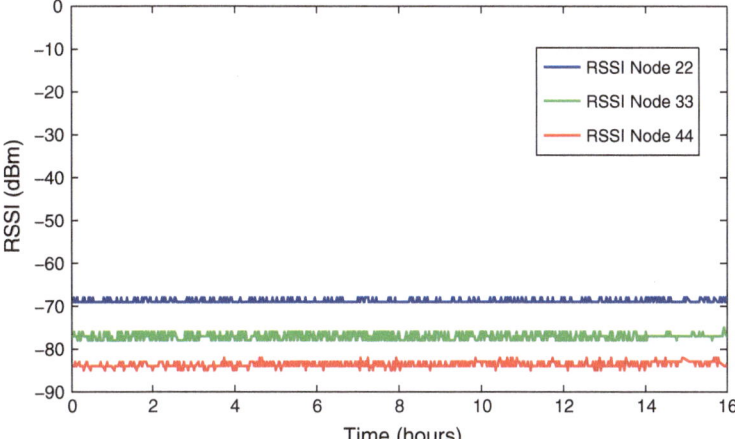

Fig. 4.7 Result of RSSI variability experiments between three different static WSN nodes and the same receiver node

communication with. The robot node transmitted these RSSI measurements to the robot using the WSN-Player interface. The robot registered the RSSI value in a RSSI map. Generation of RSSI maps was carried out by each robot individually for faster patrolling. After patrolling individually generated RSSI maps were combined. Figure 4.8-left shows the RSSI map for node $n40$ located at the top-left part of the room. Node $n43$, see Fig. 4.8-right, is located at the top-right part of the room.

4.5 Simulators for Cooperating Objects

Simulation is a necessary step before performing an experiment in a testbed. This section presents different alternatives and approaches for Cooperating Objects simulators.

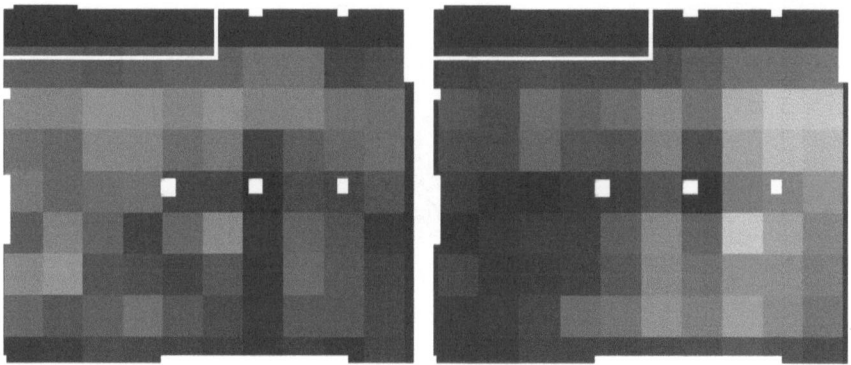

Fig. 4.8 RSSI map for node *n*40 (*left*) and *n*43 (*right*) in the CONET Integrated Testbed

4.5.1 Mobile Robot Simulators

Three open-source robot simulators have been integrated in the CONET Integrated Testbed: Stage, Gazebo and USARSim.

Stage[2] is a robot simulator belonging to the open-source Player Project [8]. It provides fairly simple, computationally-light models rather than attempting to emulate them with great fidelity. As a result Stage allows rapid prototyping of schemes for networked robots simulation. Stage is designed to support research in multi-agent autonomous systems, see Fig. 4.9-bottom-right. A wide variety of sensors and actuators models and drivers are provided for Stage as part of the Player distribution. Although it is mainly oriented towards robotics, it also provides support to easily add new elements from other technological fields. Due to its compatibility with Player, few or no changes are required to move from simulation to the hardware.

Gazebo[3] is essentially an extension of Stage for simulation in realistic environments [9]. Whereas Stage was developed to simulate large robot populations with low fidelity, Gazebo is designed to simulate small populations of robots but with high fidelity, see Fig. 4.9-bottom-left. Thus, both are complementary and are also compatible with Player: programs written for one simulator can be run on the other. Gazebo includes a rigid-body physics engine and thus generates realistic sensor feedback and physically plausible interactions between devices. Also, programs written for Gazebo can be tested in real hardware with few or no modifications.

USARSim[4] was designed as a high fidelity simulator for urban search and rescue (USAR) robots and environments [10]. Since its initial release it has been expanded to support many diverse robots and environments. High fidelity at low cost is made possible by building the simulation on top of a game engine. By off-loading the graph-

---

[2] http://playerstage.sourceforge.net

[3] http://gazebosim.org

[4] http://usarsim.sourceforge.net

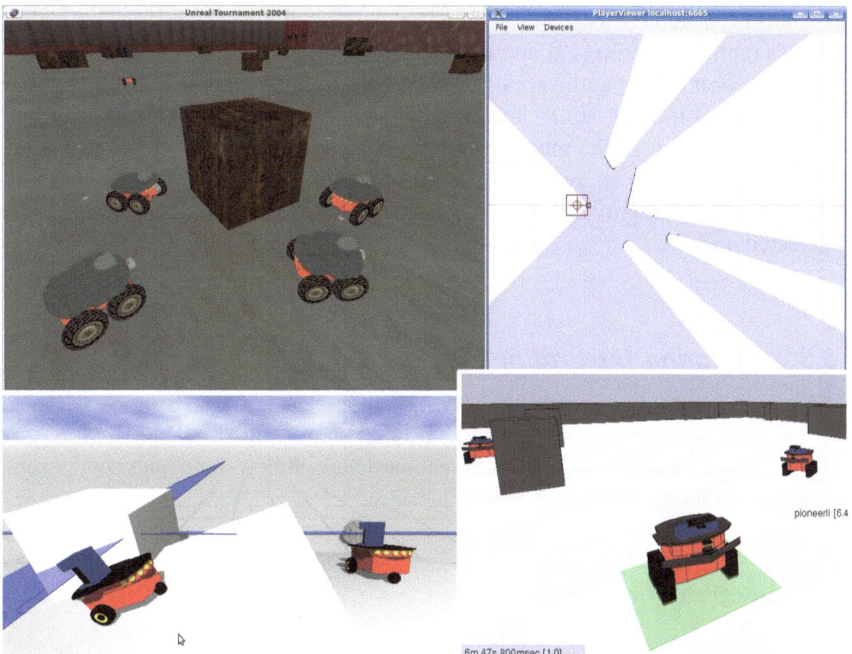

Fig. 4.9 CONET Integrated Testbed simulation using USARSim (*top-left*), PlayerViewer (*top-right*), Gazebo (*bottom-left*) and Stage (*bottom-right*)

ical aspects to a high volume commercial platform, the full effort can be devoted to the robotics-specific tasks of modeling platforms, control systems, sensors and interface tools. It can be used to provide egocentric (attached to the robot) or exocentric (third person) views of the simulation. A snapshot of a CONET Integrated Testbed USARSim simulation is shown in Fig. 4.9 top-left.

4.5.2 Sensor Network Simulators

TOSSIM [11] is an open-source discrete event simulator for TinyOS sensor networks. It uses the *nesC* programming language and the output of the *nesC* compiler is an Ansi-C file, which can either be compiled for the target platform or, in case of simulation, for the host platform. TOSSIM focuses on simulating TinyOS and its execution, rather than simulating the real world. Instead, it provides abstractions of certain real-world phenomena. With tools outside the simulation itself, users can manipulate these abstractions to implement the models they want to use in their experiments. By making complex models exterior to the simulation, TOSSIM remains flexible to the user needs and keeps the simulation simple and efficient.

Cooja [12] is a Java-based WSN simulator. It was initially designed for simulating networks running Contiki [13] operating system. Cooja can mix simulations in multiple abstraction layers: at the application level it is able to simulate the application logic developed in Java; at OS level it simulates the same code to be used in real nodes but compiled for the machine running Cooja; and at hardware level it simulates the same code to be used in real nodes. Combining the different levels in the same simulation can provide an efficient simulation as well as a detailed execution on selected nodes.

4.5.3 Simulation Integration Schemes

The software architecture of the CONET Integrated Testbed enables a variety of ways to integrate simulation tools and real hardware. Below we depict some useful integration schemes.

4.5.3.1 Integrated Simulation Tool

The most common option is the integrated simulation tool for heterogeneous Cooperating Objects. This integrated simulator is able to execute programs developed for the CONET Integrated Testbed with few or no changes. In Fig. 4.10, both mobile robots and WSN are simulated using Stage and Cooja, respectively. Player is used as integrating layer. Also, the monitoring and logging tools used in the CONET Integrated Testbed are compatible with this simulation tool.

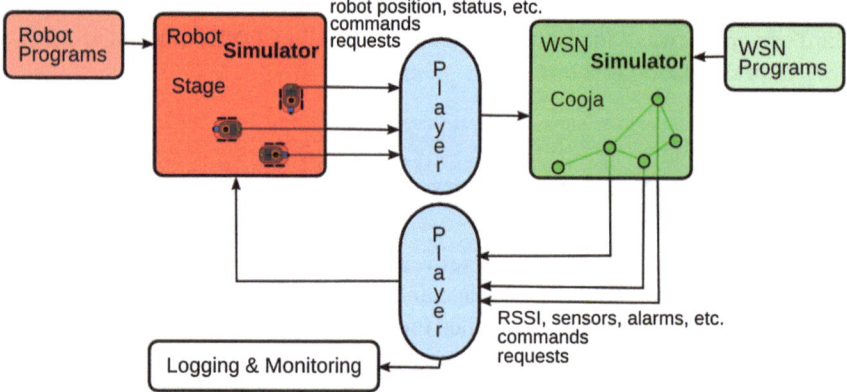

Fig. 4.10 Scheme of the simulator for the CONET Integrated Testbed

4.5.3.2 Hybrid simulation tool

The software architecture of the CONET Integrated Testbed can also be used to integrate simulators and real hardware in the same experiment, adopting a hardware-in-the-loop approach. For example, the WSN simulator could be substituted by a real WSN. The WSN sensor readings feed the robot simulator, modifying the simulated robot behavior. This scheme is particularly useful in order to experiment the performance of sensor network methods applied to the real nodes without setting the entire experiment with real hardware.

The integration of Stage with real WSN nodes has been successfully tested using this scheme. The experiment consisted of controlling the speed and orientation of a robot simulated in Stage using the measurements of the two-axis accelerometers from a real TelosB node. The readings from one axis were used to modify robot speed and those from the other axis were used to modify the robot orientation. The monitoring and logging tools of the CONET Integrated Testbed are fully compatible with this simulation tool: the logging tool registered the WSN readings as well as the simulated robot speed and orientation were remotely visualized using the CONET Integrated Testbed GUI.

The scheme is also capable of, for instance, re-playing a previously logged WSN experiment and use the data from the log as input for the simulation of robots. This scheme is useful as a way of testing the impact of real data on algorithms without setting and running a real experiment.

Figure 4.11 shows a scheme integrating a real WSN together with real robots and a further robot simulator. In this case the same robot application is executed on real hardware and simulation, which can be of interest to validate the robot performance or to evaluate robot simulation models.

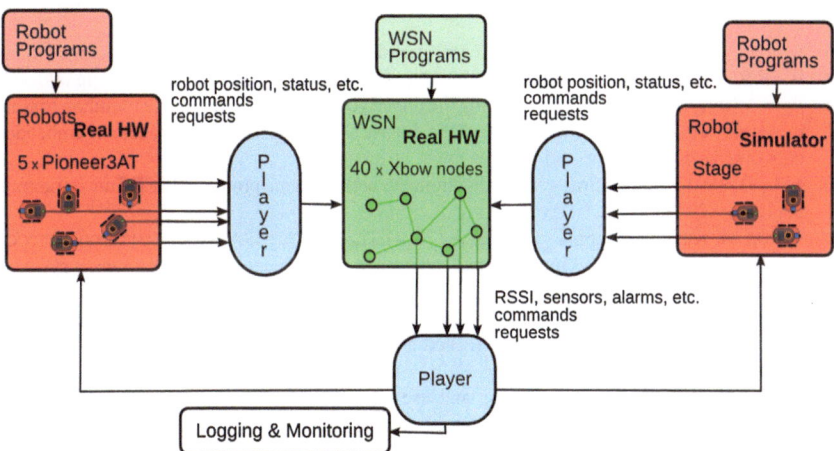

Fig. 4.11 Integrated hybrid simulator combining real experiment data and simulated data

4.6 Conclusions

As introduced in previous chapters, usability is a major issue in a testbed developed to serve a cross-section between different technological fields involving very different types of users with diverse research interests and skills. This chapter provides an insight into usability tools and basic functionalities of the CONET Integrated Testbed developed considering the requirements and guidelines presented in previous chapters.

The CONET Integrated Testbed integrates a set of basic functionalities for robots and WSN including indoors positioning, synchronization, navigation and data collection. It also includes tools to enable remote and secure testbed operation: a virtual private network, a GUI that gives full control for on-line execution and monitoring of the experiment and a website for experiment scheduling and resource allocation. Finally, different schemes and possibilities for simulating CONET Integrated Testbed experiments have been depicted. In the next chapter, some experiments that illustrate the capabilities of the CONET Integrated Testbed are briefly described.

References

1. Gnawali O, Fonseca R, Jamieson K, Moss D, Levis P (2009) Collection tree protocol. In: Proceedings of the 7th ACM conference on embedded networked sensor systems, SenSys '09, pp 1–14
2. Fox D, Burgard W, Dellaert F, Thrun S (1999) Monte carlo localization: efficient position estimation for mobile robots. In: Proceedings of the AAAI 16th conference on artificial intelligence and 11th conference on innovative applications of artificial intelligence, pp 343–349
3. Ulrich I, Borenstein J (1998) Vfh+: reliable obstacle avoidance for fast mobile robots. In: Proceedings of the IEEE international conference on robotics and automation, ICRA1998, vol 2. pp 1572–1577
4. Durham J, Bullo F (2008) Smooth nearness-diagram navigation. In: Proceedings of the IEEE/RSJ international conference on intelligent robots and systems, IROS2008, pp 690–695
5. Woo A, Tong T, Culler D (2003) Taming the underlying challenges of reliable multihop routing in sensor networks. In: Proceedings of the 1st ACM international conference on embedded networked sensor systems, pp 14–27
6. Mills D (1991) Internet time synchronization: the network time protocol. IEEE Trans Commun 39(10):1482–1493
7. Maróti M, Kusy B, Simon G, Lédeczi A (2004) The flooding time synchronization protocol. In: Proceedings of the 2nd international conference on embedded networked sensor systems, pp 39–49
8. Gerkey BP, Vaughan RT, Howard A (2003) The player/swtage project: tools for multi-robot and distributed sensor systems. In: Proceedings of the international conference on advanced robotics, pp 317–323
9. Koenig N, Howard A (2004) Design and use paradigms for gazebo, an open-source multi-robot simulator. In: IEEE/RSJ international conference on intelligent robots and systems, pp 2149–2154
10. Wang J, Lewis M, Hughes S, Koes M, Carpin S (2005) Validating usarsim for use in hri research. In: Proceedings of the human factors and ergonomics society 49th annual meeting, pp 457–461

11. Levis P, Lee N, Welsh M, Culler D (2003) Tossim: accurate and scalable simulation of entire tinyos applications. In: Proceedings of the 1st international conference on embedded networked sensor systems, SenSys '03, pp 126–137
12. Osterlind F, Dunkels A, Eriksson J, Finne N, Voigt T (2006) Cross-level sensor network simulation with cooja. In: Proceedings 2006 31st IEEE conference on local computer networks, pp 641–648
13. Dunkels A, Gronvall B, Voigt T (2004) Contiki - a lightweight and flexible operating system for tiny networked sensors. In: Proceedings of the 29th annual IEEE international conference on local computer networks, IEEE Computer Society, Washington, DC, LCN '04, pp 455–462

Chapter 5
CONET Integrated Testbed Experiments

5.1 Introduction

The CONET Integrated Testbed architecture and usability tools enable it as an experimental tool suitable for testing a wide range of experiments in a cross-section of different technological fields including mobile robots, sensor networks and robot-WSN cooperation. As pointed out in Chap. 2 preexisting Cooperating Objects testbeds lacked capabilities to hold experiments involving cooperation between elements from heterogeneous technological fields. The CONET Integrated Testbed represents a step forward in this sense, enabling full peer-to-peer interoperability, which allows it to hold a wide range of experiments that could not be performed in pre-existing testbeds.

This section briefly summarizes some of the experiments carried out in order to illustrate the capabilities of the CONET Integrated Testbed as a tool for multi-disciplinary research. The objective is not to give an exhaustive description of the performed experiments, but to show some examples in each of the main Cooperating Objects technological fields. Apart from a brief introduction, each example presents practical details of the implementation in the CONET Integrated Testbed including the modules developed and the basic functionalities that were used.

This chapter starts with Sect. 5.2, which briefly summarizes the main experiments performed since the CONET Integrated Testbed was open to the public in January 2012. Section 5.3 briefly illustrates how it can be used for experimenting localization techniques. Section 5.4 presents a set of experiments in *Range-only* Simultaneous Localization and Mapping, a fundamental problem in robotics. Section 5.5 presents two experiments that deal with explicit robot-WSN cooperation. The first one deals with dynamic robot-WSN cooperation strategies for data collection. The second one is a method for robot guiding using static WSN nodes.

J. R. Martinez-de Dios et al., *A Remote Integrated Testbed for Cooperating Objects*,
SpringerBriefs in Cooperating Objects,
DOI: 10.1007/978-3-319-01372-5_5, © The Author(s) 2014

Table 5.1 Main topics of the experiments carried out in the CONET Integrated Testbed including the main modules used

	Robots	WSN	Laser	Cameras	RSSI	Basic functions	Remote
RSSI-based localization		•			•	•	•
WSN repairing	•	•	•		•	•	
Tracking with cameras	•	•		•		•	
Multi-robot mapping	•		•			•	•
SLAM	•	•	•		•	•	
Data collection with robots	•	•	•			•	•
Multi-robot navigation	•		•			•	•
Active perception	•		•	•	•	•	
Security in WSN		•			•	•	•
Robot guiding using WSN	•	•			•	•	•
WSN routing		•				•	•
Collision avoidance	•		•	•		•	•
Optimal coverage	•	•	•			•	•

5.2 Testbed Experiment Capabilities

The CONET Integrated Testbed has been used in a wide variety of experiments. Some of them focus on mobile robots research. Others deal with sensor networks including WSN experiments and camera networks experiments. The higher number of them concentrate on robot-WSN cooperation. In some cases cooperation is exploited to improve perception. In others, robot-WSN collaboration addresses other objectives, such as increasing communication robustness. Table 5.1 briefly summarizes the main topics of the experiments that have been carried out including the main modules used in each experiment.

The CONET Integrated Testbed has been used indistinguishably to research in mobile robotics fields, in sensor networks as well as in robot-WSN experiments, allowing an unprecedented flexibility and interoperability among heterogeneous objects. Some of them have been carried out in a remote manner and others were performed on-site. A high number of basic functionalities for robots and sensor networks—described in Sect. 4.3.3—have been used releasing the user from having to program each function for each element. 30 % of the experiments were carried out remotely.

5.3 WSN Localization and Tracking Using RSSI

Localization and tracking in GPS-denied environments has attracted significant interest in the last decade. The explosion of ubiquitous systems has motivated intense research in Wireless Sensor Networks (WSN). Flexibility, reconfigurability and ease of deployment make WSN ideal for a growing number of applications. Many different target tracking methods have been developed using the signal strength of the packets interchanged between the target -assumed tagged- and the static nodes. Range-based methods, such as multilateration [1], use RSSI measurements to estimate distance to anchor nodes. Range-free methods, such as ROC-RSSI [2], rely on geometric considerations. Another approach is to learn RSSI characteristics from the environment. Fingerprinting methods, see e.g. [3], compare measurements with a previously obtained RSSI map. Other kind of cooperative perception methods have been developed between distributed static cameras at ground locations [4].

The CONET Integrated Testbed is a suitable tool for the evaluation and comparison of WSN-based localization and tracking methods. Its large indoor scenario $(500 \, \text{m}^2)$ and the combination of a network of ubiquitous sensors and mobile robots provide the elements that are usually required in localization and tracking experiments. The static WSN nodes can be used in RSSI-based methods. Bearing-based localization can be also performed using the Wireless Camera Network (WCN) comprised of CMUCam3 cameras attached to WSN nodes together with the cameras mounted on each mobile robot. Mobile robots can be used as targets ensuring motion repeatability. The robot motion can be controlled using the functionalities described in Sect. 4.3. All the measurements and variables of interest from the robot and from the WSN nodes can be logged and used for off-line processing and for experiment analysis and debugging. The testbed also registers the ground truth of the robot motion, which can be used for results analysis and evaluation. The testbed repeatability allows to perform the experiment with controlled conditions and to introduce controlled or random perturbations in order to evaluate the method robustness.

The CONET Integrated Testbed GUI allows full control of the experiment. The robot trajectory can be defined by waypoints in a preexisting file or given graphically during the experiment by clicking on the scenario. The GUI offers also the capability of logging the packets generated by the WSN and the robot ground truth location, allowing off-line processing for debugging purposes and refinement without having to repeat the experiments.

The experiment can have two main possible implementations. In a centralized scheme the RSSI measurements are transmitted to a central node, where the tracking algorithm is performed. This strategy involves energy consumption by nodes not directly involved in the sensing process and increases the probability of loosing information along the routing channels. In a semi-decentralized scheme nodes that sense the target are organized in a cluster with a cluster head (CH) that organizes and executes the tracking algorithm. Thus, the tracking information flow is kept within the cluster, simplifying the transmissions. The CONET Integrated Testbed allows implementing both schemes. Although the latter is the most frequently adopted for real implementations, centralized schemes are interesting for debugging purposes.

Figure 5.1 depicts the basic software architecture in Fig. 3.2 particularized for a typical implementation of a target tracking experiment. The Robot User Program implements basic functionalities for robot mobility control. The robot node periodi-

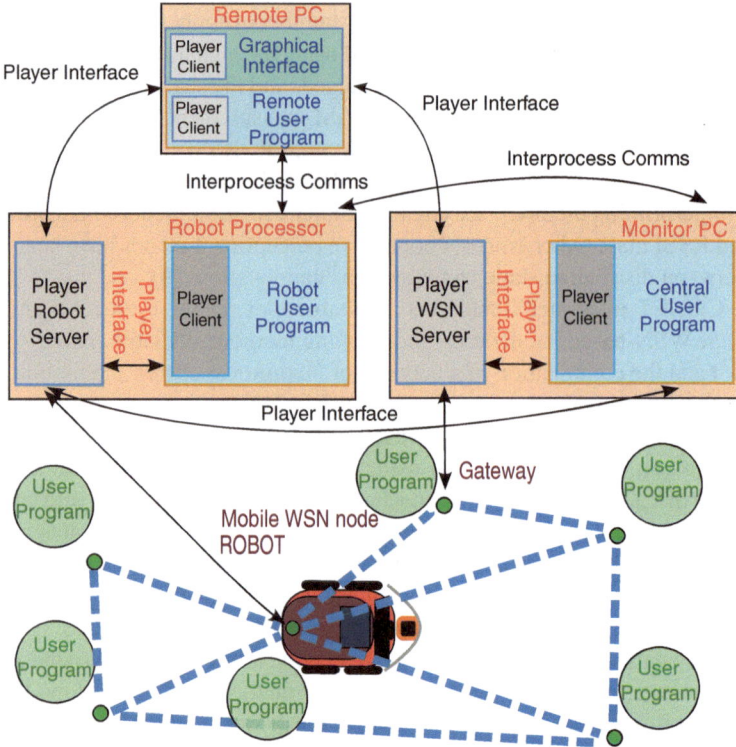

Fig. 5.1 Testbed architecture in a typical localization and tracking experiment

cally broadcasts a beacon packet. Each static WSN node implements a program that receives beacons sent by the robot node, measures its RSSI and transmits the measurement in a that value in a response packet. The robot WSN node receives the responses, extracts its content, measures the packet RSSI and computes the average between both RSSI measurements. For logging the onboard WSN node transmits to the robot the RSSI measurements and other measurements of interest through the WSN-Player interface. In some cases the onboard WSN node executes the RSSI-based tracking method program. In others, if higher computer capabilities are required, it is executed in the robot.

In some cases it is interesting to implement the tracking method in the Remote PC instead that in the Robot User Program. In this case the Remote PC connects with the Player Robot Server through the VPN and receives the RSSI measurements directly from the Player Robot Server. This allows higher experiment control and easier debugging possibilities.

The user has to implement WSN programs suitable for the experiment. The robot node program should also implement the WSN-Player communication for periodically transmitting received RSSI measurements to the robot.

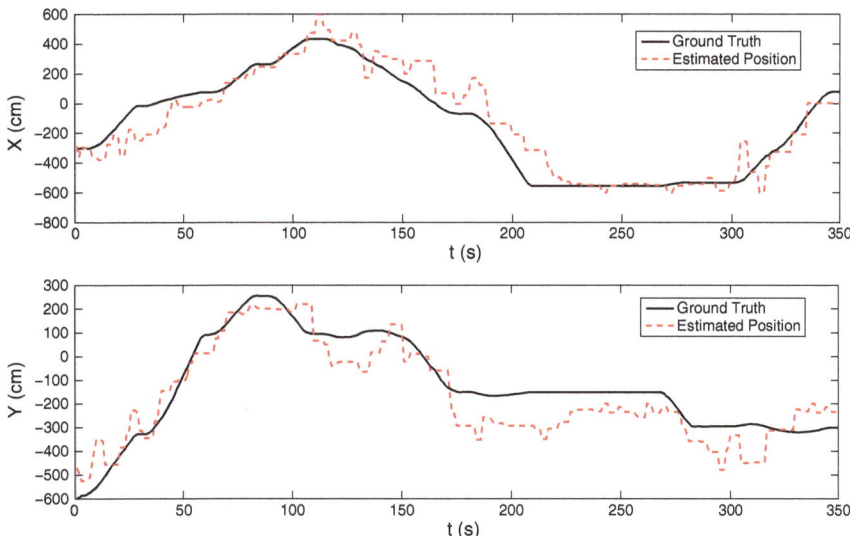

Fig. 5.2 Results of one RSSI-based object tracking experiment in axes X (*top*) and Y (*bottom*)

Different experiments of RSSI localization have been carried out in the CONET Integrated Testbed. One of the most recent implements a training based RSSI localization [5]. In indoors RSSI is highly affected by radio reflections and other interactions resulting in poor localization performance. In this experiment each static WSN node takes measurements of its neighboring static WSN nodes to compute a trained RSSI-range model that is well fitted to the peculiarities of its environment instead of using the default RSSI-range model obtained in the characterization of the testbed. The method combines the accuracy of range-based methods and the plasticity of training-based systems.

Figure 5.2 shows the results of an experiment carried out in the CONET Integrated Testbed in axes X (top) and Y (bottom). The estimated robot location is depicted in dashed red line while the ground truth is in full black line. In this experiment the Weighted Centroid Localization method (WCL) [6] is implemented in the Remote PC. In this method the mobile node i computes its location L_i with the location of static anchor nodes L_j using the following expression:

$$L_i = \frac{\sum_{j=1}^{n}(\omega_{ij} L_j)}{\sum_{j=1}^{n} \omega_{ij}}, \tag{5.1}$$

where n is the size of the data set and ω_{ij} are weighting factors that depend on the distance:

$$\omega_{ij} = \frac{1}{(D_{ij})^p}, \tag{5.2}$$

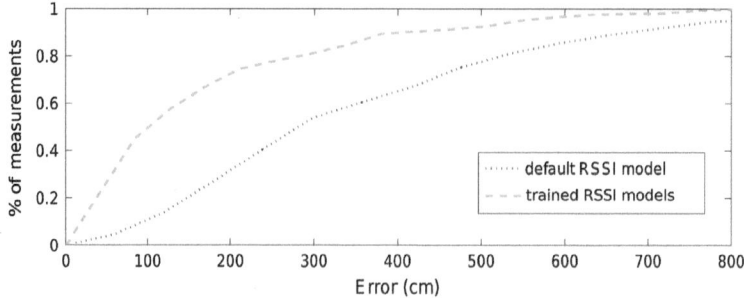

Fig. 5.3 Cumulative robot localization errors obtained by RSSI tracking using the default RSSI model and the training method used in [5]

where D_{ij} is the distance between node i and j and p is an exponent to modify the influence of distance in the weights. Higher p gives more relevance to measurements from nearby static anchor nodes. RSSI-range models become flat -i.e. insensitive-as range increases. Measurements from distant anchor nodes provide less useful information and are more affected by noise.

Figure 5.3 shows the cumulate robot localization errors obtained in the same experiment when using different methods. The cumulate error of the method that uses the default RSSI-range model is depicted in blue while the result of the method that uses the trained RSSI-range models is in green. The advantage of the trained approach is evident.

5.4 Robot Simultaneous Localization and Mapping

SLAM (Simultaneous Localization and Mapping) is a fundamental problem in robotics that has been researched for many years. In SLAM a robot placed at an unknown environment is capable of building a map without a priori knowledge and while at the same time keeping track of the robot location. It has been considered that finding solutions to this problem will make possible to build truly autonomous robots.

Many feature-based SLAM approaches use cameras as main sensors. They usually rely on methods such as SURF [7] or SIFT [8] to provide consistent feature detection and identification. *Range Only (RO)*-SLAM methods integrate range measurements between the robot and a set of static landmarks deployed in the scenario. In this case the map is given as the location of the set of landmarks.

Two basic ways to solve the initialization problem have been developed in RO-SLAM: directly introducing the measurements using a multi-hypothesis filter, e.g. [9], or performing a delayed initialization of the SLAM filter, e.g. [10]. Bearing (cameras) and range sensors allow only partial observation, i.e. it is impossible to restrict the location of one landmark with only one measurement. Thus, the SLAM methods require the robot to move and take measurements from different positions.

In the CONET Integrated Testbed the robot can be commanded to perform a predefined motion using available basic functionalities. As it moves the robot WSN node can communicate with static nodes and measure the RSSI of received packets. Then, the RSSI measurements received by the robot node are transmitted to the robot using the WSN-Player interface. The robot can integrate the measurements in the SLAM filter to simultaneously estimate the map and the robot location.

There are a number of SLAM data sets available for algorithm comparison. However, the complexity of SLAM makes it necessary to perform real tests. Authors perform experiments in their labs with different conditions hampering comparison of methods. In this sense, the CONET Integrated Testbed could be a tool for SLAM evaluation and comparison.

The implementation of SLAM experiments requires similar modules to those used in localization experiments in Sect. 5.3. The Robot User Program receives the RSSI measurements from the robot node. It implements the SLAM method, controls the robot mobility and logs all measurements and data of interest. The SLAM method can also be implemented in the Remote PC: it connects with the Player Robot Server and receives the RSSI measurements directly from it.

Mobility control is carried out using the robot basic functionalities. The user also has to implement the WSN programs suitable for the experiment. When the robot node receives a command from the robot it triggers a RSSI measurement collection event by transmitting a packet to the static nodes. The static WSN node can implement a similar program to that in Sect. 5.3.

The CONET Integrated Testbed has been used for evaluation of PF-EKF *Range Only* SLAM methods, see Fig. 5.4. These SLAM methods adopt an Extended Kalman Filter (EKF) and use auxiliary Particle Filters for landmark initialization. It is not the objective to describe these methods in detail. Only a brief description is given. The method has two distinct parts: an Extended Kalman Filter (EKF) and auxiliary Particle Filters (PFs). The EKF is the core of the estimation algorithm. It iterates prediction and update stages integrating RSSI measurements. The PF is used as an auxiliary method for initializing each of the beacon nodes.

The EKF requires two models: one to predict the robot motion and one to observe the RSSI measurements received. The prediction model adopted uses the robot kinematic model:

$$\begin{bmatrix} x_k \\ y_k \\ \theta_k \end{bmatrix} = \begin{bmatrix} x_{k-1} + T_k V_k \sin \theta_{k-1} \\ y_{k-1} + T_k V_k \cos \theta_{k-1} \\ \theta_{k-1} + T_k \alpha_k \end{bmatrix}, \qquad (5.3)$$

where (x_k, y_k, θ_k) is the robot location and heading, V_k and α_k are respectively the odometry linear and the steering velocities and T_k is the differential time between t_k and t_{k-1}.

The observation model is derived from RSSI-range models:

$$z_{k,i} = h_i(q_k) = a \log r_{k,i} + b, \qquad (5.4)$$

Fig. 5.4 Picture taken during the EKF-PF SLAM experiments carried out in the CONET Integrated Testbed

where a and b are parameters of its RSSI-range model and $r_{k,i}$ is the distance between the robot and node i at time k:

$$r_{k,i} = \sqrt{(x_k - x_{i,k})^2 + (y_k - y_{i,k})^2} \tag{5.5}$$

Both models are non-linear and the Extended version of the Kalman Filter is used.

Figure 5.5 shows the result of SLAM method in an outdoor experiment carried out in the CONET Integrated Testbed. The resulting estimated landmark position and robot trajectory are represented in red color while the ground truth values are in blue. Figure 5.6 represents the cumulative errors in the robot location with respect to the ground truth.

5.5 Robot-WSN Cooperation

Experiments involving only mobile robots and only WSN could also be carried out in the high number of robot testbeds and WSN testbeds that have been developed. The main innovation of the CONET Integrated Testbed with respect to pre-existing testbeds is the possibility of performing experiments with explicit peer-to-peer interoperability between elements from different technological fields. This section presents two exemplary experiments. In the first, robots are used to "help" WSN by collecting data from disconnected WSN nodes. In the second one, WSN nodes "help" the robots by guiding them to a given destination.

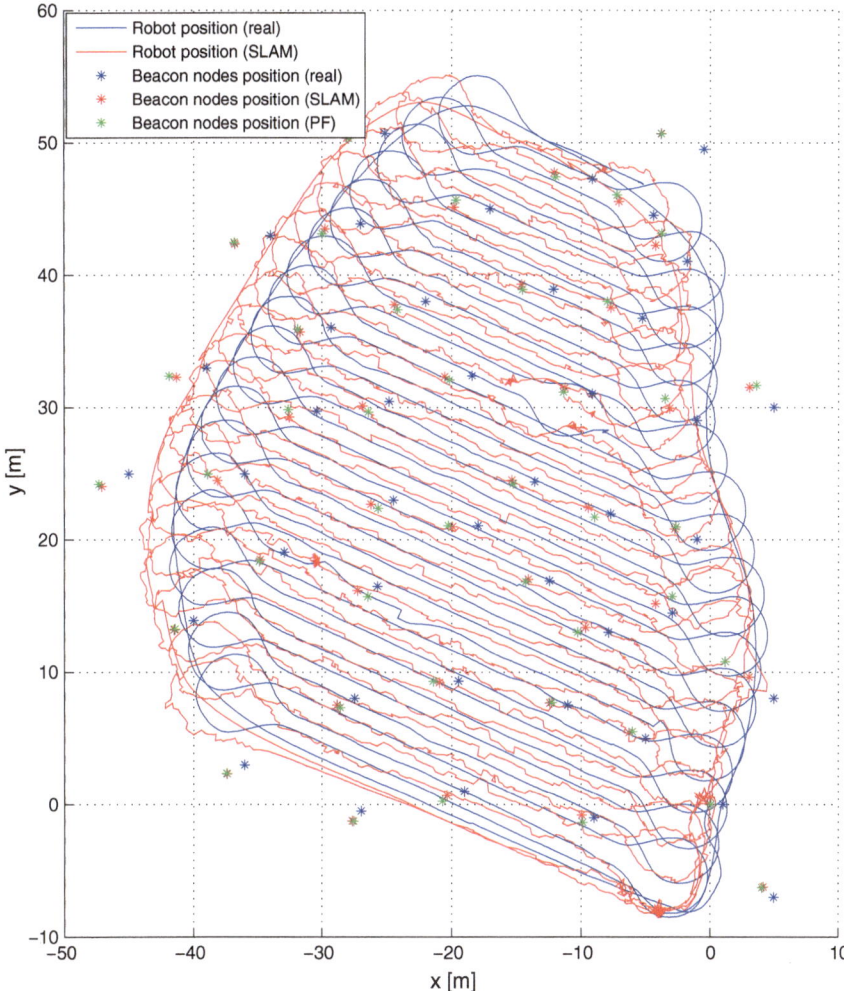

Fig. 5.5 Result of the SLAM method in an outdoor experiment

5.5.1 Robot-WSN Nodes Cooperation for Data Collection

Robots have been proposed as means for collecting data from WSN nodes scattered in an environment. The trajectories of mobile robots -collectors- equipped with communication devices can be controlled and when the collectors are close enough to the nodes, WSN data can be transferred to the robot errorless reducing transmission power and thus enlarging nodes batteries. Different data collection schemes can be distinguished depending on the management of the collected data: mobile sinks, mobile relays and mobile peers.

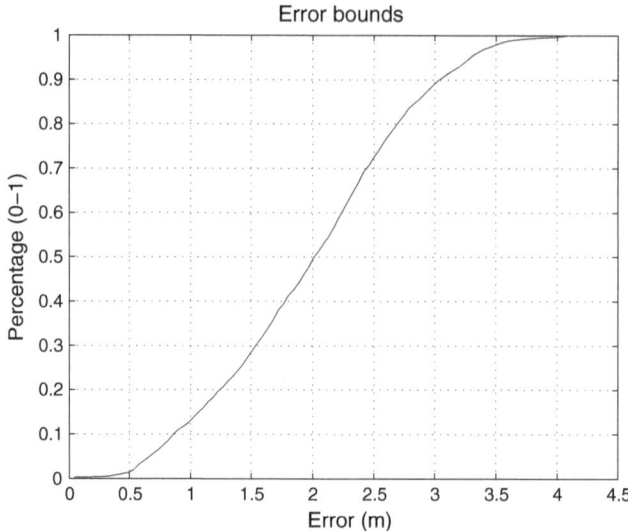

Fig. 5.6 Cumulate error of robot position estimation resulting from a SLAM experiment

In the mobile sinks approach the collectors are the destination of data originated by the WSN nodes. A number of methods have been reported. In [11] one or multiple collectors move in an area where a dense WSN is deployed to gather data generated by all nodes. In [12] people are used to collect environmental data from static sensors in a sparse network. In this case multiple collectors can be in contact with a single node at the same time.

In the mobile relay approach the robots gather data from the nodes, store them and carry the collected data to sinks or base stations. They act as mobile forwarders. One example is the data-MULE system [13, 14]. It consists of a three-tier architecture, where the middle tier is represented by relays, called Mobile Ubiquitous LAN Extensions (MULEs). Sensor nodes, which are supposed to be static, wait for a MULE to be in contact before starting communications. Then, the MULE collects data and moves to a different location, eventually passing by the base station, where the gathered data is stored and made available to remote users.

Mobile peers are ordinary mobile sensor nodes that can be both source and relay of data. Mobile peers have been applied in wildlife monitoring applications for tracking animals in the ZebraNet project [15] or in the SWIM system [16]. Mobile peers have also been applied for opportunistic data collection in urban scenarios [17].

The experiments carried out in the CONET Integrated Testbed deal with WSN data collection with Unmanned Aerial Vehicles (UAVs). In this case, the CONET Integrated Testbed is used to perform debugging tests before the field experiments.

The mobile robots take the role of the UAVs: the main objective of the experiment is to evaluate the performance of different WSN modules.

Some works have analyzed the use of aerial vehicles for WSN data collection. Most of them essentially consider WSN and UAVs as independent units that do not influence each other, lacking reactivity to unexpected events such as node failures or changes in environment conditions. In the method experimented the WSN influence the trajectory of the collector and also, the trajectory of the collector is used to modify the performance of the WSN. It requires high robot-WSN interoperability capabilities and the CONET Integrated Testbed is a suitable tool for testing such methods.

The WSN data collection method experimented is based on [18] and involves:

- dynamic autonomous organization of deployed WSN nodes in clusters using a distributed algorithm,
- dynamic generation of UAV flight plan.

Once activated by the UAV, the WSN deployed on the ground organize autonomously into clusters. The nodes in the clusters can adopt two different roles: cluster head (CH) and cluster member (CM, also non-CH). The cluster heads are responsible for coordinating the operation of the cluster members. Only CHs, one per cluster, communicates with the UAV, see Fig. 5.7.

Each CH generates a TDMA such that each non-CH node can only transmit at certain time slots and turn off the radios the rest of the time. Each non-CH periodically wakes up, takes readings from its sensors and transmits them to the CH in its TDMA slot. The CH collects all the data and aggregates it. When it receives a beacon message from the UAV it transmits the aggregated data in response. The energy consumption of CH is significantly higher than that of non-CH and a mechanism for CH rotation has been designed to select a new CH when the energy level of the current CH is below a threshold.

Fig. 5.7 Basic scheme for cluster-based UAV data collection

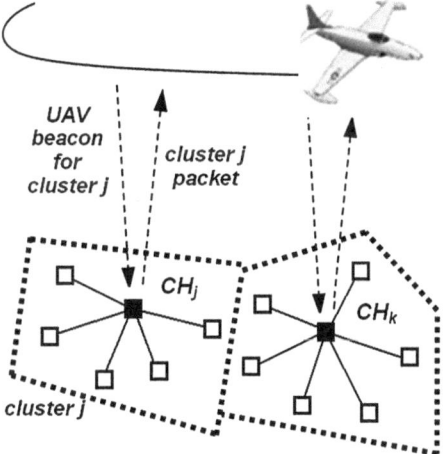

The radio transmission coverage zones of the CHs define the zones in which the UAV performs data collection and hence, the UAV flying zones. The UAV should fly on top of the CH. Since the CH role is rotated, the Base Station keeps track of which are the current CHs and commands the UAV to fly on top of them, adapting the robot motion plan to the current CHs.

The energy consumed by a CH is significantly higher than that of CM nodes. Thus, the CH role is rotated and another node in the cluster adopts the cluster head role. CH rotation selects the next CH considering the level of battery of potential candidates and also the potential change in UAV flight plan. The objective is to change the CH minimising the impact on the UAV flight plan.

Three main modules can be identified in the implementation of the method in the CONET Integrated Testbed: WSN program, the Base Station program and the robot controller, which ensures that robot trajectory passes through the waypoints commanded by the Base Station. Each WSN node includes a module for WSN cluster formation, TDMA plan generation, data aggregation, CH rotation, among others. The experiments carried out had the objective to test the performance of the WSN and of the Base Station program. It is not the objective of this section to provide details of the method. Refer to [18] for a more detailed description.

In the experiment implementation a WSN program containing the above described methods is executed in each WSN node. The Base Station program, which commands the operation and motion of the robot, is implemented as the Central User Program in the Monitor PC. It could also be implemented in the Remote User Program in the Remote PC. The robot controller is implemented locally on the processor of each robot. The experiment makes use of basic functionalities such as robot localization and obstacle avoidance. Logging functions are also employed. More than one robot could be used to speed up the collection process: each robot collects a set of WSN nodes. In this case the Base Station program should provide commands to both robots.

Figure 5.8 shows a picture taken in an WSN data collection experiment involving three robots.

5.5.2 Robot Guiding Using WSN Nodes

In the previous experiment the robots "helped" the WSN. This section describes a set of experiments in which WSN nodes "help" robots. The objective is to experimentally validate the HANSEL algorithm [19]. HANSEL is a distributed routing method for guiding mobile robots in a passive wireless environment. It assumes that the robot can detect and identify items that are in its proximity, but items cannot communicate directly with each other. These types of environments are common in some applications, such as warehouses where each product is tagged for instance with a RFID emitter. The method could be used to guide a robot that wants to find a product. For instance, HANSEL could be implemented in shopping carts.

Fig. 5.8 Picture taken in a WSN data collection experiment

The method is divided in two stages. In a learning stage, while searching for a specific node, the robots populate the memory of the static nodes they encounter with information about the nodes each robot has already found in its trajectory. Later, when a robot wants to go to a destination, it asks the static nodes, which respond with the local directions the robot should take. The method makes extensive use of the WSN-Player interface high robot-WSN interoperability.

Only two main modules are necessary to implement this experiment in the CONET Integrated Testbed: WSN program and robot program. The robots are guided by commands to move in a set of directions, typically "forward", "backward", "right" and "left". The experiment needs robot motion primitives that are currently provided by the testbed basic functionalities. The experiment also makes use of the robot localization and collision avoidance functionalities. Nodes were implemented in the experiment with WSN nodes. The node antennas were shielded and their transmission power was reduced to the minimum in order to avoid communication between WSN nodes. The experiment was executed using the CONET Integrated Testbed GUI. IP cameras provided general views of the experiment. The full video of the experiment can be seen in [20].

5.6 Conclusions

This section presents some experiments that have been carried out in the CONET Integrated Testbed. The objective is to illustrate its experimentation capabilities and provide implementation details focusing on WSN, mobile robots and robot-WSN cooperation.

The CONET Integrated Testbed allows performing mobile robots experiments, sensor network experiments and robot-WSN cooperation experiments in the same experimental tool. This enables its use to assess and compare in the same conditions solutions given to the same problem by the WSN and the robotics communities. For instance, it is possible to compare RSSI-based localization methods originated from the WSN domain, such as WCL, with localization methods originated from the robotics domain, such as fingerprinting using Particle Filters.

While WSN experiments and robot experiments could have been performed in existing WSN and robots testbeds, the presented robot-WSN cooperation experiments could not have been performed in pre-existing testbeds. The possibility to perform experiments involving explicit peer-to-peer cooperation between elements from heterogeneous technological fields is a major advantage of the CONET Integrated Testbed.

References

1. Wang X, Bischoff O, Laur R, Paul S (2009) Localization in wireless ad-hoc sensor networks using multilateration with rssi for logistic applications. Procedia Chem 1(1):461–464
2. Liu C, Wu K, He T (2004) Sensor localization with ring overlapping based on comparison of received signal strength indicator. In: International conference on mobile ad-hoc and sensor systems, pp 516–518
3. Honkavirta V, Perala T, Ali-Loytty S, Piché R (2009) A comparative survey of WLAN location fingerprinting methods. In: 6th workshop on positioning, navigation and communication, pp 243–251
4. Medeiros H, Park J, Kak A (2008) Distributed object tracking using a cluster-based kalman filter in wireless camera networks. J Sel Top Signal Process 2(4):448–463
5. De San Bernabé A, Martínez-de Dios JR, Ollero A (2012) A WSN-based tool for urban and industrial fire-fighting. Sensors 12(11):15,009–15,035
6. Blumenthal J, Grossmann R, Golatowski F, Timmermann D (2007) Weighted centroid localization in zigbee-based sensor networks. In: IEEE international symposium on intelligent signal processing, WISP 2007, pp 1–6
7. Murillo A, Guerrero J, Sagues C (2007) Surf features for efficient robot localization with omnidirectional images. In: Proceedings of IEEE international conference on robotics and automation, pp 3901–3907, April 2007.doi:10.1109/ROBOT.2007.364077
8. Se S, Lowe D, Little J (2001) Vision-based mobile robot localization and mapping using scale-invariant features. In: Proceedings of the IEEE international conference on robotics and automation, vol 2, pp 2051–2058. doi:10.1109/ROBOT.2001.932909
9. Blanco JL, Fernandez-Madrigal JA, Gonzalez J (2008) Efficient probabilistic range-only slam. In: Proceedings of IEEE/RSJ international conference on intelligent robots and systems, pp 1017–1022. doi:10.1109/IROS.2008.4650650
10. Djugash J, Singh S, Kantor G, Zhang W (2006) Range-only slam for robots operating cooperatively with sensor networks. In: Proceedings of the 2006 IEEE international conference on robotics and automation, pp 2078–2084. doi:10.1109/ROBOT.2006.1642011
11. Rao J, Wu T, Biswas S (2008) Network-assisted sink navigation protocols for data harvesting in sensor networks. In: IEEE Wireless communications and networking conference, WCNC 2008. IEEE, pp 2887–2892
12. Anastasi G, Borgia E, Conti M, Gregori E (2011) A hybrid adaptive protocol for reliable data delivery in wsns with multiple mobile sinks. Comput J 54(2):213–229

13. Shah RC, Roy S, Jain S, Brunette W (2003) Data mules: modeling a three-tier architecture for sparse sensor networks. In: Proceedings of the 1st IEEE 2003 international workshop on sensor network protocols and applications, IEEE, pp 30–41

14. Jain S, Shah RC, Brunette W, Borriello G, Roy S (2006) Exploiting mobility for energy efficient data collection in wireless sensor networks. Mob Netw Appl 11(3):327–339

15. Juang P, Oki H, Wang Y, Martonosi M, Peh LS, Rubenstein D (2002) Energy-efficient computing for wildlife tracking: Design tradeoffs and early experiences with zebranet. In: ACM sigplan notices, ACM, vol 37, pp 96–107

16. Small T, Haas ZJ (2003) The shared wireless infostation model: a new ad hoc networking paradigm (or where there is a whale, there is a way). In: Proceedings of the 4th ACM international symposium on mobile ad hoc networking & computing, ACM, pp 233–244

17. Campbell AT, Eisenman SB, Lane ND, Miluzzo E, Peterson RA (2006) People-centric urban sensing. In: Proceedings of the 2nd annual international workshop on wireless internet, ACM, p 18

18. Martinez-de Dios J, Lferd K, Núnez G, Torres-Gonzalez A, Ollero A (2013) Cooperation between uas and wireless sensor networks for efficient data collection in large environments. J Intell Robot Syst 70(1–4):491–508. doi:10.1007/s10846-012-9733-2

19. Zuniga M, Hauswirth M (2009) Hansel: distributed localization in passive wireless environments. In: 6th annual IEEE communications society conference on sensor, mesh and ad hoc communications and networks, 2009, SECON'09, IEEE, pp 1–9

20. Zuniga M, Jiménez-González A, Martínez-de Dios J. Video of HANSEL experiment in the CONET integrated testbed. http://www.youtube.com/watch?v=uxJXOSO7qrM Accessed in June 2013

Chapter 6
Conclusions

This book presented the CONET Integrated Testbed. Developed for the evaluation and comparison of Cooperating Objects methods and techniques, this experimental tool is intended to fill a gap in existing testbeds, which have low capabilities to enable cooperation between different technological fields.

The CONET Integrated Testbed has been developed to allow full equanimity between heterogeneous elements from different Cooperating Objects fields, enabling the possibility to allocate an unprecedented range of experiments. It employs a modular and flexible architecture based on Player. From a hardware perspective the testbed comprises the elements and sensors more frequently used in Cooperating Objects experiments: a set of mobile robots equipped with a laser range finder and a RGB-D camera, among others, a camera network and a WSN with static and mobile nodes of different models and manufacturers.

The testbed has been developed as an easy-to-use tool and includes a set of basic functionalities straightforward usable by the users in their experiments, a GUI to facilitate secure remote interface with the testbed during the experiment and a website for resource allocation and scheduling.

With the CONET Integrated Testbed it is possible to perform sensor networks experiments, networked robots experiments and robot-WSN cooperation experiments in the same experimental tool, which allows assessing and comparing in the same conditions solutions to the same problem originated in different Cooperating Objects technological fields.

The CONET Integrated testbed can be considered the first general-purpose Cooperating Object testbed. However, upgrading and updating efforts should keep on driven by the feedback from testbed users and the analysis of the technological and scientific tendencies.

A proportion of users experimented reluctance to adopt a new tool to which they have to get used. Experiment support tools are the best arguments against this first barrier. From this point of view, analyzing users feedback we have noticed that basic functionalities, straightforward usable by testbed users, play an important role. The range of basic functionalities have been widened since the opening of the

J. R. Martinez-de Dios et al., *A Remote Integrated Testbed for Cooperating Objects*, 75
SpringerBriefs in Cooperating Objects,
DOI: 10.1007/978-3-319-01372-5_6, © The Author(s) 2014

CONET Integrated Testbed, following users suggestions. Our objective is to keep incorporating more and more basic functionalities and users support modules. Below are some that will be added in the short term:

- advanced multi-robot navigation methods for autonomous cooperative surveillance,
- multi-robot task allocation methods,
- vision-based methods for localization and mapping,
- laser-based mapping,
- support for other WSN operating systems apart from TinyOS and Contiki,
- security functions for WSN,
- support functions for interprocess communication.

The CONET Integrated Testbed has been equipped with the most widely used devices and sensors in networked robotics, WSN and robot-WSN experiments. This integrating effort should keep on in the next years. Of particular interest is the integration of Unmanned Aerial Systems such as quad-copters. Highly suitable for indoor testbeds, they allow extending experiments and envisioned applications from 2D to 3D with high potential scientific and industrial impact. The addition of quad-copters would not require updating the current architecture but would involve components for more accurate measurement of UAV pose. In this sense the installation of the VICON system[1] in the testbed room is under analysis.

In the recent years Robotic Operating Systems (ROS) has gained increasing interest. A large and active community is working in ROS not only in robotics domains. Also, there are efforts to integrate WSN into ROS. Migration of the CONET Integrated Testbed to ROS would allow benefiting for such a large community in terms of software and documentation upgrading and support for new robots, sensors and other devices. Also, user programs developed for ROS will be straightforward integrated in a testbed based on ROS, improving the testbed usability. Being Player one of the main predecessor of ROS, there are many commonalities between both, and migration of the architecture of the CONET Integrated Testbed to ROS should not require high efforts.

As pointed out in Chap. 2 one of the main trends in Cooperating Objects testbeds is the integration of different testbeds with heterogeneous resources and capabilities in larger testbed structures and federations. Federated testbeds share resources virtually or physically, which for instance allows testing the same experiment in different settings. In some cases, the integration consists of using the same API. In others, such as the EU-funded *FP7 Future Internet Research and Experimentation* (*FIRE*) program,[2] the integration is done by sharing a common basic architecture. The integration of the CONET Integrated Testbed in larger testbed federations, and particularly in *FIRE*, is object of current analysis.

---

[1] http://www.vicon.com

[2] EC (2013) FP7 Future Internet Research and Experimentation (FIRE). http://www.ict-fire.eu/home/fire-projects.html

One important future research line is to take Cooperating Objects testbeds out of the laboratories and put them in the real application. Systems that are used in the real world are, at the same time, used as testbeds. These real-world testbeds enable testing in real operational conditions, which is a relevant advantage, and are also efficient in terms of cost. Very few real-world testbeds exist and many research topics are still to be explored. Mechanisms to ensure harmonious coexistence between the testbed experiments and the real system performance is a main issue.

Index

J. R. Martinez-de Dios et al., *A Remote Integrated Testbed for Cooperating Objects*, 79
SpringerBriefs in Cooperating Objects,
DOI: 10.1007/978-3-319-01372-5, © The Author(s) 2014